The author presents a basic introduction to the world of genetic engineering. Some 20 years have passed since the first recombinant DNA molecules were constructed at Stanford University. Genetic engineering is now a reality and this book simply and concisely describes to the student the full range of enabling technologies available. The book takes the reader from basic molecular biology through to chapters dealing with the principles behind working with nucleic acids, together with cloning strategies and the tools of the trade. The author discusses the applications of genetic engineering in a clear and engaging manner.

The book is essential reading for first and second-year undergraduates, as well as being of interest to sixth-form students and their teachers. Medical students and general practitioners will also find this book useful for background information.

An introduction to genetic engineering

The Institute of Biology aims to advance both the science and practice of biology. Besides providing the general editors for this series, the Institute publishes two journals *Biologist* and *the Journal of Biological Education*, conducts examinations, arranges national and local meetings and represents the views of its members to government and other bodies. The emphasis of the *Studies in Biology* will be on subjects covering major parts of first-year undergraduate courses. We will be publishing new editions of the 'bestsellers' as well as publishing additional new titles.

In preparation for the *Studies in Biology*

Photosynthesis, 5th edition, D.O. Hall and K.K. Rao

An introduction to genetic engineering

Desmond S.T. Nicholl
Senior lecturer,
Department of Biological Sciences
University of Paisley

Published in association with the Institute of Biology

Published by the Press Syndicate of the University of Cambridge
The Pitt Building, Trumpington Street, Cambridge CB2 1RP
40 West 20th Street, New York, NY 10011–4211, USA
10 Stamford Road, Oakleigh, Melbourne 3166, Australia

First published 1994
Reprinted 1994

Printed in Great Britain at the University Press, Cambridge

A catalogue record for this book is available from the British Library

Library of Congress cataloguing in publication data

Nicholl, Desmond S. T.
An introduction to genetic engineering / Desmond S.T. Nicholl.
p. cm. – (Studies in biology)
"Published in association with the Institute of Biology."
Includes bibliographical references (p.) and index.
ISBN 0-521-43054-2 (hc). – ISBN 0-521-43634-6 (pb)
1. Genetic engineering. I. Institute of Biology. II. Title.
III. Series.
QH442.N53 1994
575.1'0724—dc20 93-8176 CIP

ISBN 0 521 43054 2 hardback
ISBN 0 521 43634 6 paperback

SE

Contents

General preface to the series

Charged by its Royal Charter to promote biology and its understanding, the Institute of Biology recognises that it is not possible for any one text book to cover the entirety of a course. If evidence was needed, the success of the *Studies in Biology* series was a testimony to the need for specialist, up-to-date publications in education. The Institute is therefore pleased to collaborate with Cambridge University Press in producing a new title in the *Studies in Biology* series.

The new series is set to provide as great a boon to the new generation of students as the original did to their parents.

Suggestions and comments from readers will always be welcomed and should be addressed either to the Studies in Biology Editorial Board at Cambridge University Press or c/o The Books Committee at the Institute.

Robert Priestley
The General Secretary

The Institute of Biology
20 Queensberry Place
London SW7 2DZ

Preface

It is now some 20 years since the first recombinant DNA molecules were constructed at Stanford University and genetic engineering became a reality. During these two decades the impact of recombinant DNA technology has been felt worldwide, in many diverse disciplines, and the potential for future development seems almost limitless.

The aim of this book is to provide a basic technical introduction to the subject of genetic engineering. It is designed to complement, and not compete with, the many excellent texts already available. Examples have been chosen to illustrate the underlying principles of the technology, and to outline some of its applications. Genetic engineering is often seen as a form of scientific sorcery, when in fact it is merely an enabling technology that has opened up the world of the gene in a spectacular way. I hope that this book may help to remove some of the mystery that surrounds the subject.

My thanks go to Professor Bill Stevely and to Drs Gordon Bickerstaff, Peter Birch and Simon Hettle for their invaluable comments on the manuscript. Any errors of fact or interpretation that remain are, of course, my own. I am also grateful to my wife and family for their patience and support during a task that inevitably proved more difficult and time-consuming than expected.

This book is dedicated to my parents.

D.S.T. Nicholl
Paisley

1

Introduction

1.1 What is genetic engineering?

Progress in any scientific discipline is dependent on the availability of techniques and methods that extend the range and sophistication of experiments which may be performed. Over the last 20 years or so this has been demonstrated in spectacular fashion by the emergence of genetic engineering. This field has grown rapidly to the point where, in many laboratories around the world, it is now routine practice to isolate a specific DNA fragment from the genome of an organism, determine its base sequence, and assess its function. What is particularly striking is that this technology is readily accessible by individual scientists, without the need for large-scale equipment or resources outside the scope of a reasonably well-found research laboratory.

The term **genetic engineering** is often thought to be rather emotive or even trivial, yet it is probably the label that most people would recognise. However, there are several other terms which may be used to describe the technology, including **gene manipulation**, **gene cloning**, **recombinant DNA technology**, **genetic modification**, and the **new genetics**. There are also legal definitions used in administering regulatory mechanisms in countries where genetic engineering is practised.

Although there are many diverse and complex techniques involved, the basic principles of genetic manipulation are reasonably simple. The premise on which the technology is based is that genetic information, encoded by DNA and arranged in the form of genes, is a resource which can be manipulated in various ways to achieve certain goals in both pure and

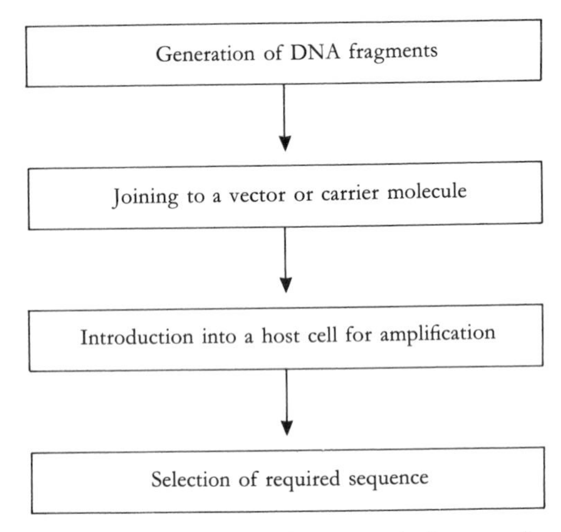

Fig. 1.1. The four steps in a gene cloning experiment.

applied science and medicine. There are three main areas in which genetic manipulation is of value:

- Basic research on gene structure and function
- Production of useful proteins by novel methods
- Generation of transgenic plants and animals

In later chapters I look at some of the ways in which genetic manipulation has contributed to these research areas.

The mainstay of genetic manipulation is the ability to isolate a single DNA sequence from the genome. This is the essence of **gene cloning**, and can be considered as a series of four steps (Fig. 1.1). Successful completion of these steps provides the genetic engineer with a specific DNA sequence, which may then be used for a variety of purposes. A useful analogy is to consider gene cloning as a form of **molecular agriculture**, enabling the production of large amounts (in genetic engineering this means micrograms or milligrams) of a particular DNA sequence.

One aspect of the new genetics that has given cause for concern is the debate surrounding the potential applications of the technology. The term **genethics** has recently been coined to describe the ethical problems that exist in modern genetics, which are likely to increase in both number and complexity as genetic engineering technology becomes more sophisticated. The use of transgenic plants and animals, investigation of the human genome, gene therapy, and many other topics are of concern not just to the scientist but to the population as a whole. It remains to be seen if we can use

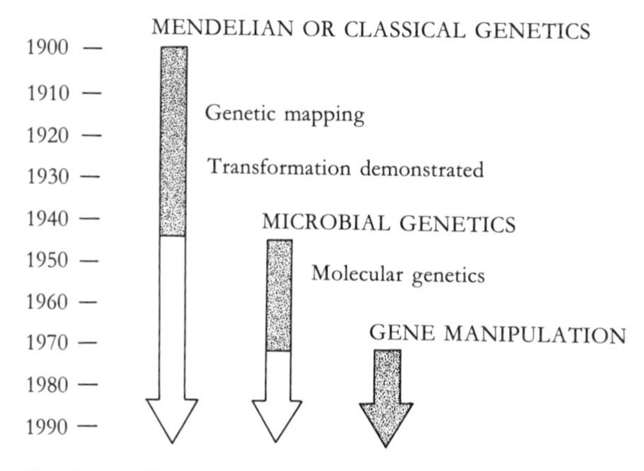

Fig. 1.2. The history of genetics since 1900. Shaded areas represent the periods of major development in each branch of the subject.

genetic engineering for the overall benefit of mankind, and avoid the misuse of technology which often accompanies scientific achievement.

1.2 Laying the foundations

Although the techniques used in gene manipulation are relatively new, it should be remembered that development of these techniques was dependent on the knowledge and expertise provided by microbial geneticists. We can consider the development of genetics as falling into three main eras (Fig. 1.2). The science of genetics really began with the rediscovery of Gregor Mendel's work at the turn of the century, and the next 40 years or so saw the elucidation of the principles of inheritance and genetic mapping. Microbial genetics emerged in the mid 1940s, and the role of DNA as the genetic material was firmly established. During this period great advances were made in understanding the mechanisms of gene transfer between bacteria, and a broad knowledge base was established from which later developments would emerge.

The discovery of the structure of DNA by James Watson and Francis Crick in 1953 provided the stimulus for the development of genetics at the molecular level, and the next few years saw a period of intense activity and excitement as the main features of the gene and its expression were determined. This work culminated with the establishment of the complete genetic code in 1966 – the stage was now set for the appearance of the new genetics.

1.3 First steps

In the late 1960s there was a sense of frustration among scientists working in the field of molecular biology. Research had developed to the point where progress was being hampered by technical constraints, as the elegant experiments that had helped to decipher the genetic code could not be extended to investigate the gene in more detail. However, a number of developments provided the necessary stimulus for gene manipulation to become a reality. In 1967 the enzyme **DNA ligase** was isolated. This enzyme can join two strands of DNA together, a prerequisite for the construction of recombinant molecules, and can be regarded as a sort of molecular glue. This was followed by the isolation of the first **restriction enzyme** in 1970, a major milestone in the development of genetic engineering. Restriction enzymes are essentially molecular scissors, which cut DNA at precisely defined sequences. Such enzymes can be used to produce fragments of DNA that are suitable for joining to other fragments. Thus, by 1970, the basic tools required for the construction of recombinant DNA were available.

The first recombinant DNA molecules were generated at Stanford University in 1972, utilising the cleavage properties of restriction enzymes (scissors) and the ability of DNA ligase to join DNA strands together (glue). The importance of these first tentative experiments cannot be overestimated. Scientists could now join different DNA molecules together, and could link the DNA of one organism to that of a completely different organism. The methodology was extended in 1973 by joining DNA fragments to the plasmid pSC101, which is an **extrachromosomal element** isolated from the bacterium *Escherichia coli*. These recombinant molecules behaved as **replicons**, i.e. they could replicate when introduced into *E. coli* cells. Thus, by creating recombinant molecules *in vitro*, and placing the construct in a bacterial cell where it could replicate *in vivo*, specific fragments of DNA could be isolated from bacterial colonies that formed clones (colonies formed from a single cell, in which all cells are identical) when grown on agar plates. This development marked the emergence of the technology which became known as **gene cloning** (Fig. 1.3).

The discoveries of 1972 and 1973 triggered off what is perhaps the biggest scientific revolution of all – the new genetics. The use of the new technology spread very quickly, and a sense of urgency and excitement prevailed. This was dampened somewhat by the realisation that the new

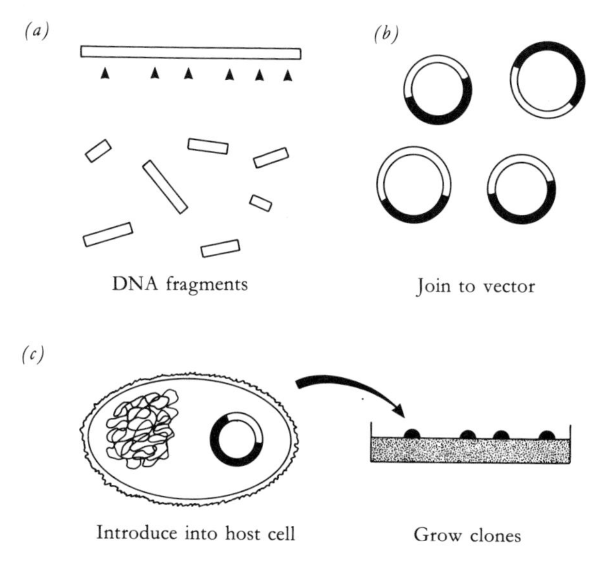

Fig. 1.3. Cloning DNA fragments. (*a*) The DNA is fragmented into smaller pieces. (*b*) The fragments are then joined to a carrier molecule or vector to produce recombinant DNA molecules. (*c*) These are then introduced into a host cell for propagation as individual clones.

technology could give rise to potentially harmful organisms, exhibiting undesirable characteristics. It is to the credit of the biological community that measures were adopted to regulate the use of gene manipulation, and that progress in contentious areas was limited until more information became available regarding the possible consequences of the inadvertent release of organisms containing recombinant DNA.

1.4 What is in store?

The remaining chapters fall into three main areas. Chapter 2 (Basic molecular biology) and Chapter 3 (Working with nucleic acids) provide background information about DNA and the techniques used when working with it. Chapter 4 (The tools of the trade) and Chapter 5 (The biology of genetic engineering) outline the essential enzymes and biological systems that are required for genetic engineering. In the final part of the book, Chapters 6, 7 and 8 deal with the applications of genetic engineering techniques. Chapter 6 (Cloning strategies) outlines the various protocols that may be used to clone DNA. Chapter 7 (Selection, screening and analysis of recombinants) describes how particular DNA sequences can be selected from collections of cloned fragments, and Chapter 8 (Genetic engineering in action) illustrates the applications of the technology with reference to specific examples.

At the end of each chapter a **concept map** is given, covering the main points of the chapter. Concept mapping is a technique that can be used to structure information and provide links between various topics. The concept maps provided here are essentially summaries of the chapters, and may be examined either before or after reading the chapter. I hope that they prove to be a useful addition to the text.

Some suggestions for further reading are given at the end of the book. No reference has been made to the primary literature, as this is well documented in the books and articles mentioned in this section. A glossary of terms used has also been provided; this may be particularly useful for readers who may be unfamiliar with the terminology used in molecular biology.

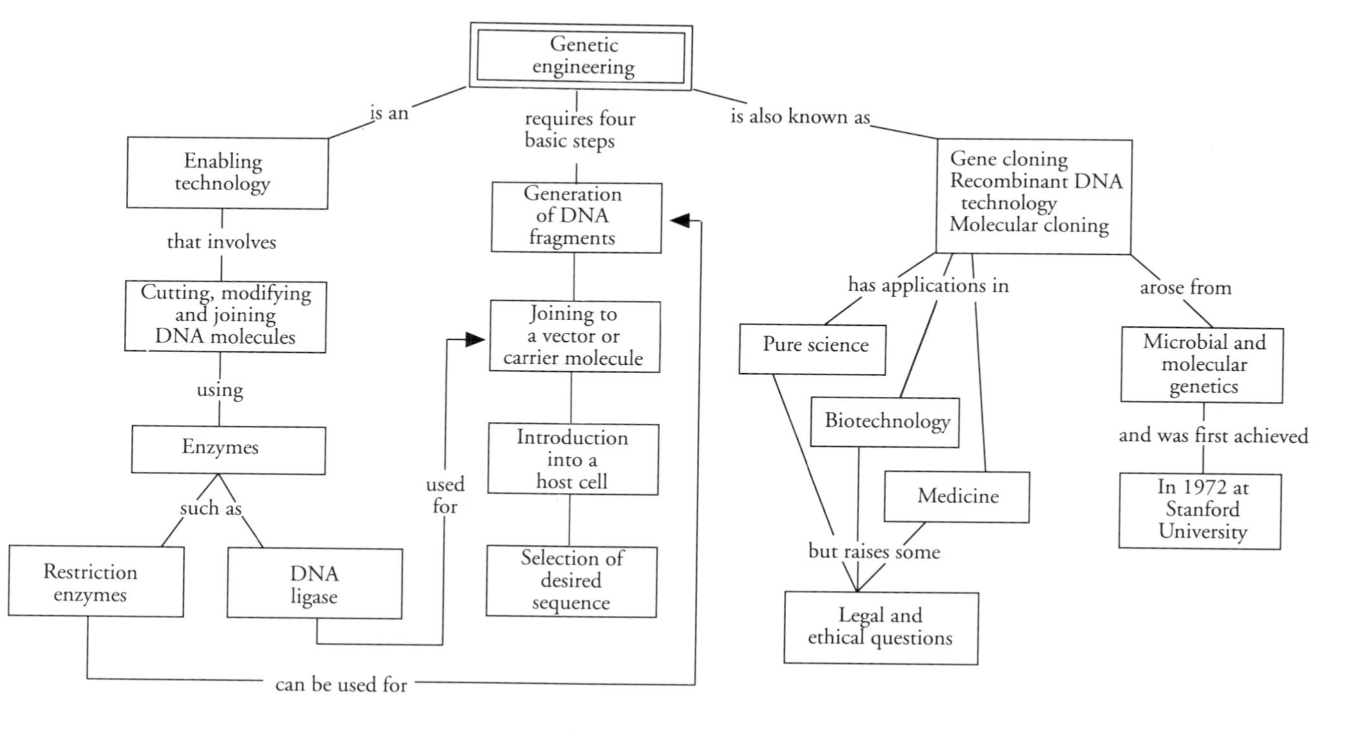

Concept map 1.

2

Basic molecular biology

In this chapter I present a brief overview of the structure and function of DNA. This provides the non-specialist reader with an introduction to the topic, and may also act as a useful refresher for those who have some background knowledge of DNA. More extensive accounts of the topics presented here may be found in the textbooks listed in Suggestions for further reading.

2.1 The flow of genetic information

It is a remarkable fact that an organism's characteristics are encoded by a four-letter alphabet, defining a language of three-letter words. The letters of this alphabet are the bases adenine (A), guanine (G), cytosine (C) and thymine (T), with triplet combinations of these bases making up the 'dictionary' that is the genetic code.

The expression of genetic information is achieved ultimately *via* proteins, particularly the **enzymes** that catalyse the reactions of metabolism. Proteins are condensation **heteropolymers** derived from amino acids, of which 20 are used in natural proteins. Given that a protein may consist of several hundred amino acid residues, the number of different proteins that may be made is essentially unlimited; thus great diversity of protein form and function can be achieved using an elegantly simple coding system. The genetic code is shown in Table 2.1.

The flow of genetic information is unidirectional, from DNA to protein, with **messenger RNA** (mRNA) as an intermediate. The copying of DNA-

Table 2.1. *The genetic code*

First base (5′-end)	Second base U	C	A	G	Third base (3′-end)
U	Phe	Ser	Tyr	Cys	U
	Phe	Ser	Tyr	Cys	C
	Leu	Ser	STOP	STOP	A
	Leu	Ser	STOP	Trp	G
C	Leu	Pro	His	Arg	U
	Leu	Pro	His	Arg	C
	Leu	Pro	Gln	Arg	A
	Leu	Pro	Gln	Arg	G
A	Ile	Thr	Asn	Ser	U
	Ile	Thr	Asn	Ser	C
	Ile	Thr	Lys	Arg	A
	Met	Thr	Lys	Arg	G
G	Val	Ala	Asp	Gly	U
	Val	Ala	Asp	Gly	C
	Val	Ala	Glu	Gly	A
	Val	Ala	Glu	Gly	G

Note: Uracil (U) is found in RNA in place of thymine, and is used in this Table. Codons read 5′→3′, thus AUG specifies Met. The three-letter abbreviations for the amino acids are as follows: Ala, Alanine; Arg, Arginine; Asn, Asparagine; Asp, Aspartic acid; Cys, Cysteine; Gln, Glutamine; Glu, Glutamic acid; Gly, Glycine; His, Histidine; Ile, Isoleucine; Leu, Leucine; Lys, Lysine; Met, Methionine; Phe, Phenylalanine; Pro, Proline; Ser, Serine; Thr, Threonine; Trp, Tryptophan; Tyr, Tyrosine; Val, Valine. The three codons UAA, UAG and UGA specify no amino acid and terminate translation.

encoded genetic information into RNA is known as **transcription** (T_C), with the further conversion into protein being termed **translation** (T_L). This concept of information flow is known as the **Central Dogma** of molecular biology, and is an underlying theme in all studies on gene expression.

A further two aspects of information flow may be added to this basic model to complete the picture. Firstly, duplication of the genetic material prior to cell division represents a DNA–DNA transfer, and is known as DNA **replication**. A second addition, with important consequences for the genetic engineer, stems from the fact that some viruses have RNA instead

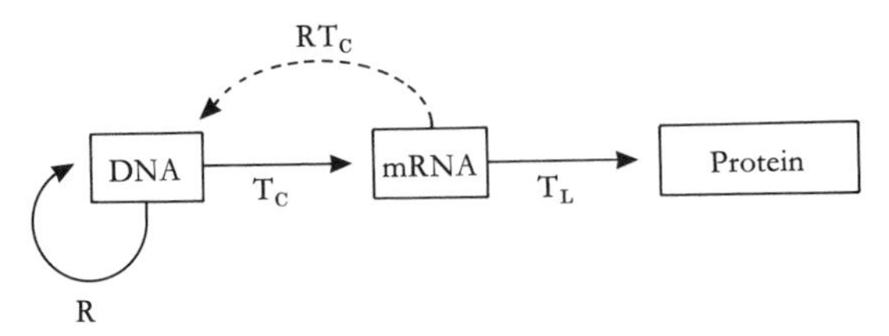

Fig. 2.1. The Central Dogma. This states that information flow is unidirectional, from DNA to mRNA to protein. The processes of transcription (T_C), translation (T_L) and DNA replication (R) obey this rule. An exception is found in some RNA viruses, which carry out a process known as reverse transcription (RT_C), producing a DNA copy of their viral, RNA genome.

of DNA as their genetic material. These viruses (chiefly members of the retrovirus group) have an enzyme called **reverse transcriptase** (an RNA-dependent DNA polymerase) which produces a double-stranded DNA molecule from the single-stranded RNA genome. Thus in these cases the flow of genetic information is reversed with respect to the normal convention. The Central Dogma is summarised in Fig. 2.1.

2.2 The structure of DNA and RNA

In most organisms, the primary genetic material is double-stranded DNA. What is required of this molecule? Firstly, it has to be **stable**, as genetic information may need to function in a living organism for up to 100 years or more. Secondly, the molecule must be capable of **replication**, to permit dissemination of genetic information as new cells are formed during growth and development. Thirdly, there should be the potential for limited alteration to the genetic material (**mutation**), to enable evolutionary pressures to exert their effects. The DNA molecule fulfils these criteria of stability, replicability and mutability, and when considered with RNA provides an excellent example of the premise that 'structure determines function'.

Nucleic acids are heteropolymers composed of monomers known as **nucleotides**; a nucleic acid chain is therefore often called a **polynucleotide**. The monomers are themselves made up of three components: a sugar, a phosphate group, and a nitrogenous base. The two types of nucleic acid (DNA and RNA) are named according to the sugar component of the nucleotide, with DNA having 2′-deoxyribose as the sugar (hence **D**eoxyribo**N**ucleic**A**cid) and RNA having ribose (hence **R**ibo**N**ucleic**A**cid). The

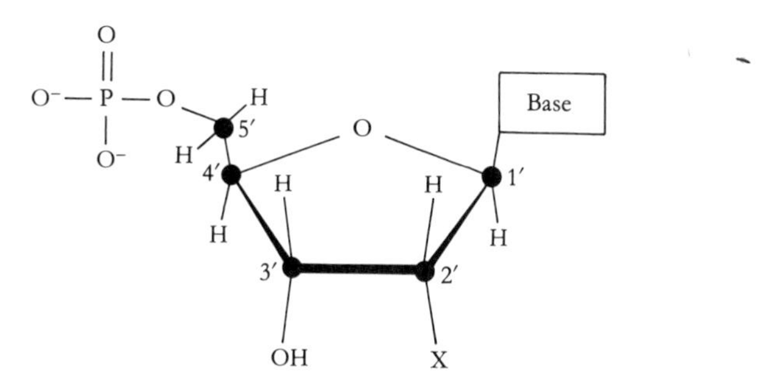

Fig. 2.2. The structure of a nucleotide. Carbon atoms are represented by solid circles, numbered 1′ to 5′. In DNA the sugar is deoxyribose, with a hydrogen atom at position X. In RNA the sugar is ribose, which has a hydroxyl group at position X. The base can be A,G,C or T in DNA, and A,G,C or U in RNA.

sugar/phosphate components of a nucleotide are important in determining the structural characteristics of polynucleotides, with the nitrogenous bases determining their information storage and transmission characteristics. The structure of a nucleotide is summarised in Fig. 2.2.

Nucleotides can be joined together by a 5′–3′ phosphodiester linkage, which confers directionality on the polynucleotide. Thus the 5′ end of the molecule will have a free phosphate group, and the 3′ end a free hydroxyl group; this has important consequences for the structure, function and manipulation of nucleic acids. In a double-stranded molecule such as DNA, the sugar–phosphate chains are found in an **antiparallel** arrangement, with the two strands running in different directions.

The nitrogenous bases are the important components of nucleic acids in terms of their coding function. In DNA the bases are as listed in section 2.1 above, namely adenine (A), guanine (G), cytosine (C) and thymine (T). In RNA the base thymine is replaced by uracil (U), which is functionally equivalent. Chemically adenine and guanine are **purines**, which have a double ring structure, whereas cytosine and thymine (and uracil) are **pyrimidines**, which have a single ring structure. In DNA the bases are paired, A with T and G with C. This pairing is determined both by the bonding arrangements of the atoms in the bases, and by the spatial constraints of the DNA molecule, the only satisfactory arrangement being a purine:pyrimidine base-pair. The bases are held together by hydrogen bonds, two in the case of an A·T base-pair and three in the case of a G·C base-pair. The structure and base pairing arrangement of the four DNA bases is shown in Fig. 2.3.

The DNA molecule *in vivo* usually exists as a right-handed double helix called the *B*-form. This is the structure proposed by Watson and Crick in

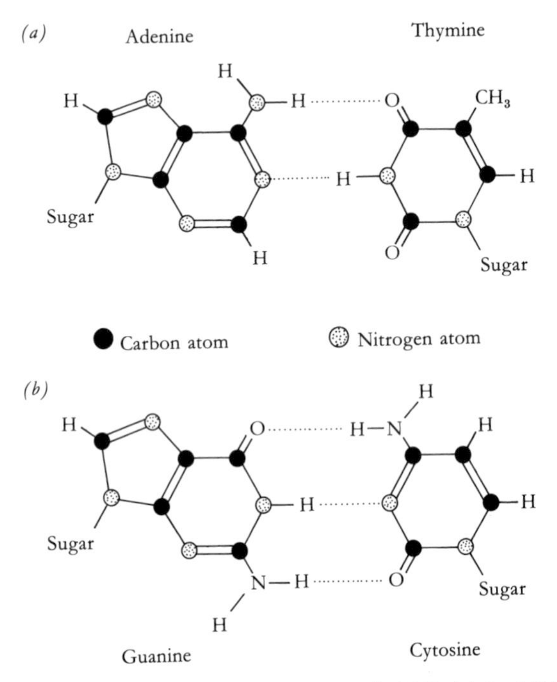

Fig. 2.3. Base-pairing arrangements in DNA. (*a*) An A·T base-pair is shown. The bases are linked by two hydrogen bonds (dotted lines). (*b*) A G·C base-pair, with three hydrogen bonds.

1953. Alternative forms of DNA include the *A*-form (right-handed helix) and the *Z*-form (left-handed helix). Although DNA structure is a complex topic, particularly when the higher-order arrangements of DNA are considered, a simple representation will suffice here, as shown in Fig. 2.4.

The structure of RNA is similar to that of DNA, the main chemical differences being the presence of ribose instead of 2′-deoxyribose and uracil instead of thymine. RNA is also most commonly single-stranded, although short stretches of double-stranded RNA may be found in self-complementary regions. There are three main types of RNA molecule found in cells: **messenger** RNA (mRNA), **ribosomal** RNA (rRNA) and **transfer** RNA (tRNA). Ribosomal RNA is the most abundant class of RNA molecule, making up some 85% of total cellular RNA. It is associated with **ribosomes**, which are an essential part of the translational machinery. Transfer RNAs make up about 10% of total RNA, and provide the essential specificity that enables the insertion of the correct amino acid into the protein that is being synthesised. Messenger RNA, as the name suggests, acts as the carrier of genetic information from the DNA to the translational machinery, and is usually less than 5% of total cellular RNA.

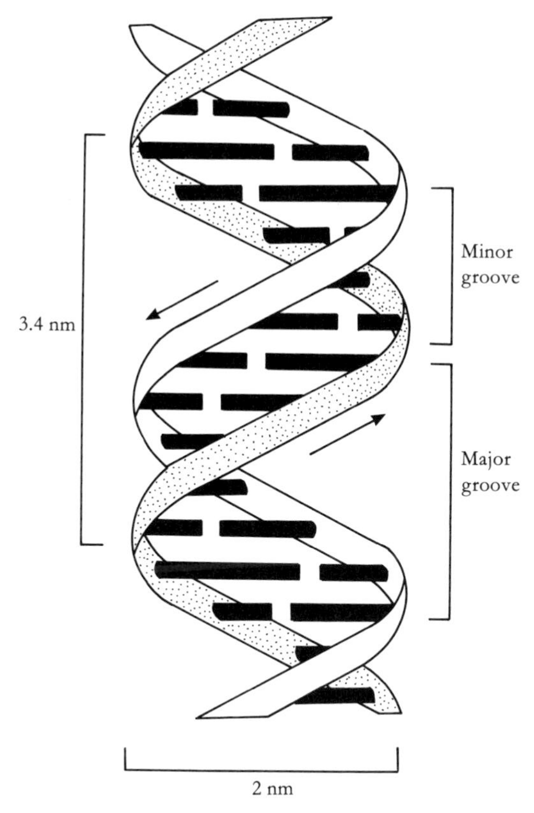

Fig. 2.4. The double helix. This is DNA in the commonly found *B*-form. The right-handed helix has a diameter of 2 nm and a pitch of 3.4 nm, with 10 base-pairs per turn. The sugar–phosphate 'backbones' are antiparallel (arrowed) with respect to their 5′→3′ orientations. One of the sugar–phosphate chains has been shaded for clarity. The purine–pyrimidine base-pairs are formed across the axis of the helix.

2.3 Gene organisation

The gene can be considered as the basic unit of genetic information. Genes have been studied since the turn of the century, when genetics became established. Before the advent of molecular biology and the realisation that genes were made of DNA, study of the gene was largely indirect; the effects of genes were observed in phenotypes and the 'behaviour' of genes was analysed. Despite the apparent limitations of this approach, a vast amount of information about how genes functioned was obtained, and the basic tenets of transmission genetics were formulated.

As the gene was studied in greater detail, the terminology associated with this area of genetics became more extensive, and the ideas about genes were modified to take account of developments. The term **gene** is usually taken to represent the genetic information transcribed into a single RNA molecule, which is in turn translated into a single protein. Exceptions are genes for rRNA and tRNA species, which are not translated. Genes are located on **chromosomes**, and the region of the chromosome where a particular gene is found is called the **locus** of that gene. In diploid organisms, which have their chromosomes arranged as homologous pairs, different forms of the same gene are known as **alleles**.

The double-stranded DNA molecule has the potential to store genetic information in either strand, although in most organisms only one strand is used to encode any particular gene. There is some degree of confusion over the nomenclature of the two DNA strands, which may be called coding/non-coding, sense/antisense, plus/minus, transcribed/non-transcribed or template/non-template. In some cases different authors use the same terms in different ways, which adds to the confusion. Current recommendations from the International Union of Biochemistry (IUB) and the International Union of Pure and Applied Chemistry (IUPAC) favour the terms coding/non-coding, with the **coding** strand of DNA taken to be the **mRNA-like** strand. This convention will be used in this book where coding function is specified. The terms **template** and **non-template** will be used to describe DNA strands when there is not necessarily any coding function involved, as in the copying of DNA strands during cloning procedures. Thus genetic information is expressed by transcription of the **non-coding** strand of DNA, which produces an mRNA molecule that has the same sequence as the coding strand of DNA (although the RNA has uracil substituted for thymine, see Fig. 2.8(*a*)). The sequence of the coding strand is usually reported when dealing with DNA sequence data, as this permits easy reference to the sequence of the RNA.

In addition to the sequence of bases that specifies the codons in a protein-coding gene, there are other important regulatory sequences associated with genes (Fig. 2.5). A site for starting transcription is required, and this encompasses a region which binds RNA polymerase, known as the **promoter**, and a specific start point for transcription. A stop site for transcription is also required. From T_C start to t_C stop is sometimes called the **transcriptional unit**, i.e. the DNA region that is copied into RNA. Within this transcriptional unit there may be regulatory sites for translation, namely a T_L start and a t_L stop signal. Other sequences involved in the control of gene expression may be present either upstream or downstream from the gene itself.

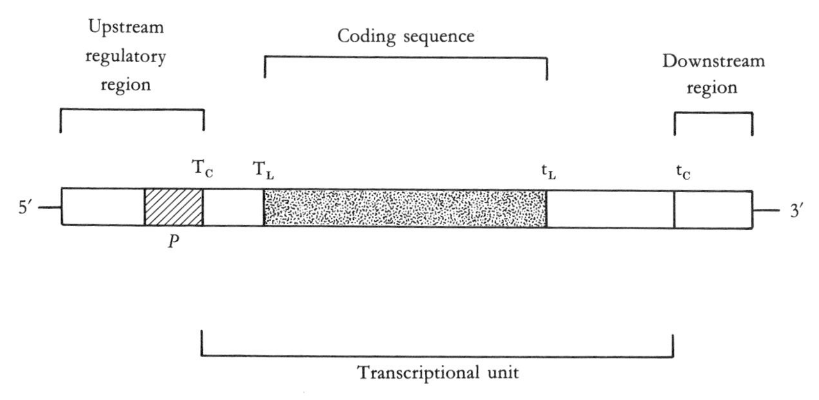

Fig. 2.5. Gene organisation. The transcriptional unit produces the RNA molecule, and is defined by the transcription start site (T_C) and stop site (t_C). Within the transcriptional unit lies the coding sequence, from the translation start site (T_L) to the stop site (t_L). The upstream regulatory region may have controlling elements such as enhancers or operators in addition to the promoter (*P*), which is the RNA polymerase-binding site.

2.3.1 Gene structure in prokaryotes

In prokaryotic cells such as bacteria, genes are usually found grouped together in **operons**. The operon is a cluster of genes that are related (often coding for enzymes in a metabolic pathway), and which are under the control of a single promoter/regulatory region. Perhaps the best known example of this arrangement is the *lac* operon (Fig. 2.6), which codes for the enzymes responsible for lactose catabolism. Within the operon there are three genes that code for proteins (termed **structural** genes) and an upstream control region encompassing the promoter and a regulatory site called the **operator**. In this control region there is also a site which binds a complex of cyclic AMP and CRP (cyclic AMP receptor protein), which is important in positive regulation (stimulation) of transcription. Lying outside the operon itself is the repressor gene, which codes for a protein (the Lac repressor) that binds to the operator site and is responsible for negative control of the operon by blocking the binding of RNA polymerase.

The fact that structural genes in prokaryotes are often grouped together means that the transcribed mRNA may contain information for more than one protein. Such a molecule is known as a **polycistronic** mRNA. Thus much of the genetic information in bacteria is expressed *via* polycistronic mRNAs whose synthesis is regulated in accordance with the needs of the cell at any given time. This system is flexible and efficient, and enables the cell to adapt quickly to changing environmental conditions.

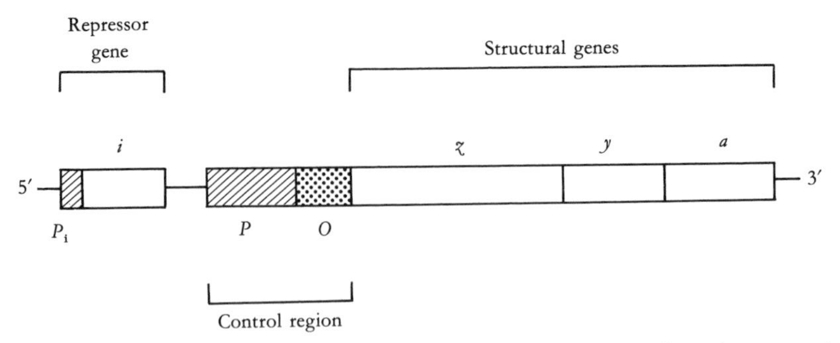

Fig. 2.6. The *lac* operon. The structural genes *lacZ, lacY* and *lacA* (noted as *z, y* and *a*) encode ß-galactosidase, galactoside permease and a transacetylase, respectively. The cluster is controlled by a promoter (*P*) and an operator region (*O*). The operator is the binding site for the repressor protein, encoded by the *lacI* gene (*i*). The repressor gene is controlled by its own promoter, P_i.

2.3.2 Gene structure in eukaryotes

Eukaryotic cells have a membrane-bound nucleus, which means that the genetic information stored in DNA is separated from the site of translation, which is in the cytoplasm. The picture is complicated further by the presence of genetic information in mitochondria and chloroplasts, which have their own separate genomes that specify many of the components required by these organelles. This compartmentalisation has important consequences for regulation, both genetic and metabolic, and thus gene structure and function in eukaryotes is more complex than in prokaryotes.

The most startling discovery concerning eukaryotic genes was made in 1977, when it became clear that eukaryotic genes contained ‘extra’ pieces of DNA that did not appear in the mRNA that the gene encoded. These sequences are known as **intervening sequences** or **introns**, with the sequences that will make up the mRNA being called **exons**. In some cases the number and total length of the introns exceeds that of the exons, as in the chicken ovalbumin gene, which has a total of seven introns making up more than 75% of the gene.

The presence of introns obviously has important implications for the expression of genetic information in eukaryotes, in that the introns must be removed before the mRNA can be translated. This is carried out in the nucleus, where the introns are spliced out of the primary transcript. Further intranuclear modification includes the addition of a ‘cap’ at the 5′ terminus and a ‘tail’ of adenine residues at the 3′ terminus. These modifications are part of what is known as RNA processing, and the end product is a fully

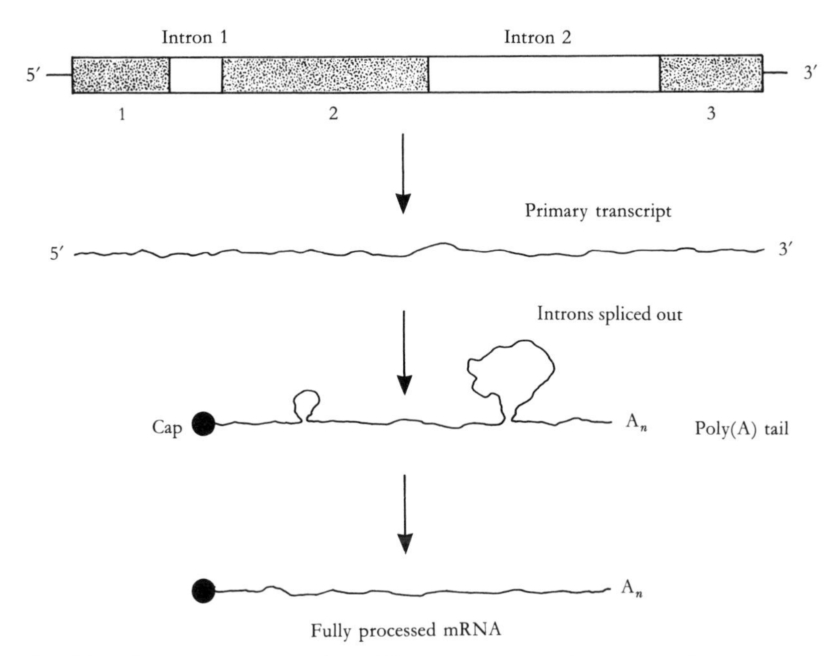

Fig. 2.7. Structure and expression of the mammalian ß-globin gene. The gene contains two intervening sequences or introns. The expressed sequences (exons) are shaded and numbered. The primary transcript is processed by capping, polyadenylation and splicing to yield the fully functional mRNA.

functional mRNA that is ready for export to the cytoplasm for translation. The structures of the mammalian ß-globin gene and its processed mRNA are outlined in Fig. 2.7 to illustrate eukaryotic gene structure.

2.4 Gene expression

As shown in Fig. 2.1, the flow of genetic information is from DNA to protein. Whilst a detailed knowledge of gene expression is not required in order to understand the principles of genetic engineering, it is important to be familiar with the basic features of transcription and translation. A brief description of these processes is given here.

Transcription involves synthesis of an RNA from the DNA template provided by the non-coding strand of the transcriptional unit in question. The enzyme responsible is **RNA polymerase** (DNA-dependent RNA polymerase). In prokaryotes there is a single RNA polymerase enzyme, but in eukaryotes there are three types of RNA polymerase (I, II and III). These

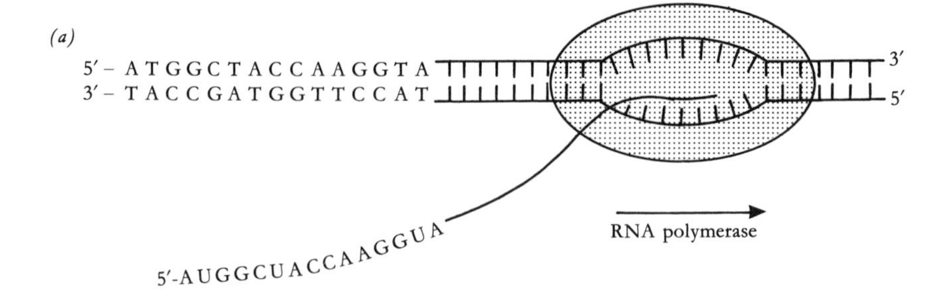

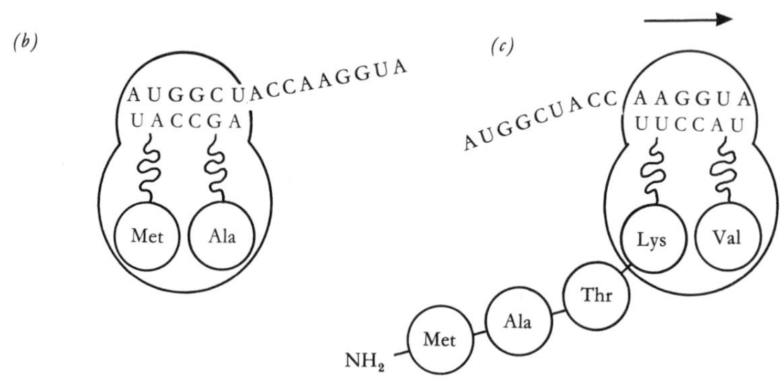

Fig. 2.8. Transcription and translation. (*a*) Transcription involves synthesis of mRNA by RNA polymerase. Part of the DNA/mRNA sequence is given. The mRNA has the same sequence as the coding strand in the DNA (the non-template strand), apart from U being substituted for T. (*b*) The mRNA is being translated. The amino acid residue is inserted into the protein in response to the codon/anticodon recognition event in the ribosome. The first amino acid residue is encoded by AUG in the mRNA (tRNA anticodon UAC), which specifies methionine (see Table 2.1 for the genetic code). (*c*) The ribosome has translated the remainder of the sequence in a similar way. The ribosome translates the mRNA in a 5′→3′ direction, with the polypeptide growing from its N terminus. The residues in the polypeptide chain are joined together by peptide bonds.

synthesise ribosomal, messenger and transfer/5 S ribosomal RNAs respectively. All RNA polymerases are large multisubunit proteins with relative molecular masses of around 500 000.

Transcription has several component stages, these being (i) DNA/RNA polymerase binding, (ii) chain initiation, (iii) chain elongation, and (iv) chain termination and release of the RNA. Promoter structure is important in determining the binding of RNA polymerase, but will not be dealt with here. When the RNA molecule is released, it may be immediately available for translation (as in prokaryotes) or it may be processed and exported to the cytoplasm (as in eukaryotes) before translation occurs.

Translation requires an mRNA molecule, a supply of charged tRNAs (tRNA molecules with their associated amino acid residues) and ribosomes (composed of rRNA and ribosomal proteins). The ribosomes are the sites where protein synthesis occurs; in prokaryotes, ribosomes are composed of three rRNAs and some 52 different ribosomal proteins. The ribosome is a complex structure that essentially acts as a 'jig' which holds the mRNA in place so that the **codons** may be matched up with the appropriate **anticodon** on the tRNA, thus ensuring that the correct amino acid is inserted into the growing polypeptide chain. The mRNA molecule is translated in a 5′→3′ direction, corresponding to polypeptide elongation from N terminus to C terminus.

Although transcription and translation are complex processes, the essential features (with respect to information flow) may be summarised as shown in Fig. 2.8. In conjunction with the brief descriptions presented above, this should provide enough background information about gene structure and expression to enable subsequent sections of the text to be linked to these processes where necessary.

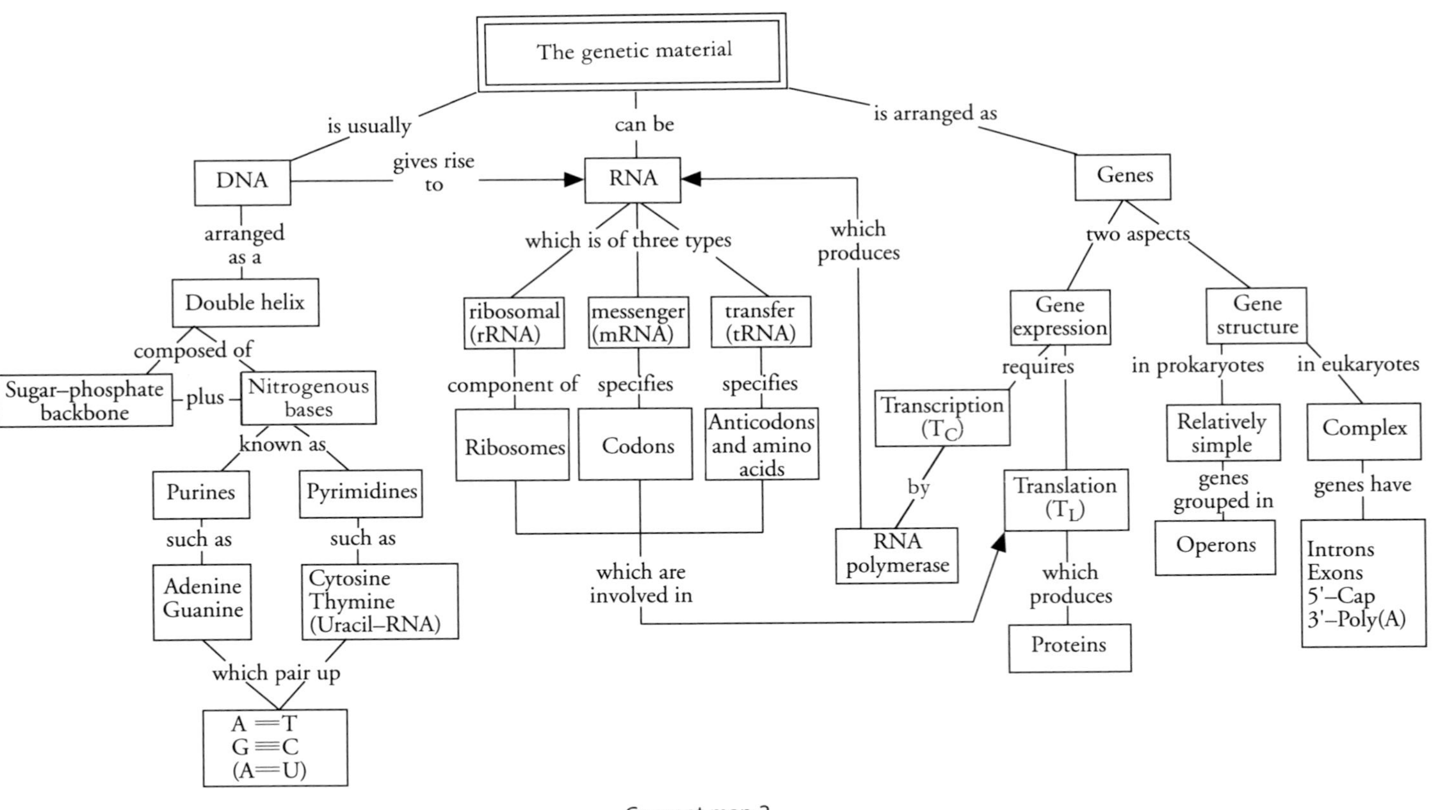

Concept map 2.

3

Working with nucleic acids

Before examining some of the specific techniques used in gene manipulation, it is useful to consider the basic methods required for handling, quantifying and analysing nucleic acid molecules. It is often difficult to make the link between theoretical and practical aspects of a subject, and an appreciation of the methods used in routine work with nucleic acids may be of help when the more detailed techniques of gene cloning and analysis are described.

3.1 Isolation of DNA and RNA

Every gene manipulation experiment requires a source of nucleic acid, in the form of either DNA or RNA. It is therefore important that reliable methods are available for isolating these components from cells. The first step in any isolation protocol is disruption of the starting material, which may be viral, bacterial, plant or animal. The method used to open cells should be as gentle as possible, preferably utilising enzymatic degradation of cell wall material (if present) and detergent lysis of cell membranes. If more vigorous methods of cell disruption are required (as is the case with some types of plant cell material), there is the danger of mechanically shearing large DNA molecules, and this can hamper the production of representative recombinant molecules during subsequent processing.

Following cell disruption, most methods involve a deproteinisation stage. This is often achieved by one or more extractions using phenol or phenol/chloroform mixtures. On the formation of an emulsion and

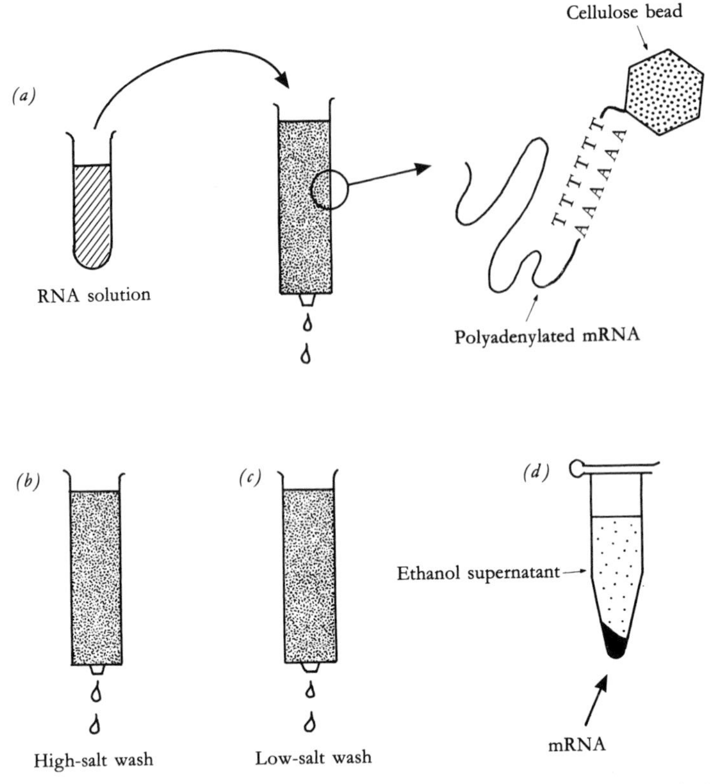

Fig. 3.1. Preparation of mRNA by affinity chromatography using oligo(dT)-cellulose. (*a*) Total RNA in solution is passed through the column in a high-salt buffer, and the oligo(dT) tracts bind the poly(A) tails of the mRNA. (*b*) Residual RNA is washed away with high-salt buffer, and (*c*) the mRNA is eluted by washing with a low salt buffer. (*d*) The mRNA is then precipitated under ethanol and collected by centrifugation.

subsequent centrifugation to separate the phases, protein molecules partition into the phenol phase and accumulate at the interface. The nucleic acids remain mostly in the upper aqueous phase, and may be precipitated from solution using isopropanol or ethanol (see section 3.2).

If a DNA preparation is required, the enzyme **ribonuclease** (RNase) can be used to digest the RNA in the preparation. If mRNA is needed for cDNA synthesis, a further purification can be performed by using oligo(dT)-cellulose to bind the poly(A) tails of eukaryotic mRNAs (Fig. 3.1). This gives substantial enrichment for mRNA and enables most contaminating DNA, rRNA and tRNA to be removed.

The technique of gradient centrifugation is often used to prepare DNA, particularly plasmid DNA (pDNA). In this technique a caesium chloride solution containing the DNA preparation is spun at high speed in an ultracentrifuge. Over a long period (up to 48 h in some cases) a density gradient is formed and the pDNA forms a band at one position in the centrifuge tube. The band may be taken off and the CsCl removed by dialysis to give a pure preparation of pDNA.

3.2 Handling and quantification of nucleic acids

It is often necessary to use very small amounts of nucleic acid (typically **micro-**, **nano-** or **picograms**) during a cloning experiment. It is obviously impossible to handle these amounts directly, so most of the measurements that are done involve the use of aqueous solutions of DNA and RNA. The concentration of a solution of nucleic acid can be determined by measuring the absorbance at 260 nanometres, using a spectrophotometer. An A_{260} of 1.0 is equivalent to a concentration of 50 $\mu g\ ml^{-1}$ for double-stranded DNA, or 40 $\mu g\ ml^{-1}$ for single-stranded DNA or RNA. If the A_{280} is also determined, the A_{260}/A_{280} ratio indicates if there are contaminants present, such as residual phenol or protein. The A_{260}/A_{280} ratio should be 1.8 for pure DNA and 2.0 for pure RNA preparations.

In addition to spectrophotometric methods, the concentration of DNA may be estimated by monitoring the fluorescence of bound **ethidium bromide**. This dye binds between the DNA bases (intercalates) and fluoresces orange when illuminated with ultraviolet (u.v.) light. By comparing the fluorescence of the sample with that of a series of standards, an estimate of the concentration may be obtained. This method can detect as little as 1–5 ng of DNA, and may be used when u.v.-absorbing contaminants make spectrophotometric measurements impossible. Having determined the concentration of a solution of nucleic acid, any amount (in theory) may be dispensed by taking the appropriate volume of solution. In this way nanogram or picogram amounts may be dispensed with reasonable accuracy.

Precipitation of nucleic acids is an essential technique that is used in a variety of applications. The two most commonly used precipitants are isopropanol and ethanol, ethanol being the preferred choice for most applications. When added to a DNA solution in a ratio, by volume, of 2:1 in the presence of 0.2 M salt, ethanol causes the nucleic acids to come out of solution. Although it used to be thought that low temperatures (-20 °C or -70 °C) were necessary, this is not an absolute requirement, and 0 °C appears to be adequate. After precipitation the nucleic acid can be recovered by centrifugation, which causes a pellet of nucleic acid material to form at the bottom of the tube. The pellet can be dried and the nucleic acid resuspended in the buffer appropriate to the next stage of the experiment.

3.3 Radiolabelling of nucleic acids

A major problem encountered in any cloning procedure is that of keeping track of the small amounts of nucleic acid involved. This problem is magnified at each stage of the process, because losses mean that the amount of material usually diminishes after each step. One way of tracing the material is to label the nucleic acid with a radioactive molecule (usually a deoxynucleoside triphosphate (dNTP), labelled with ^{3}H or ^{32}P), so that portions of each reaction may be counted in a scintillation counter to determine the amount of nucleic acid present.

A second application of radiolabelling is in the production of highly radioactive nucleic acid molecules for use in hybridisation experiments. Such molecules are known as radioactive **probes**, and have a variety of uses (see sections 3.4 and 7.2). The difference between labelling for tracing purposes and labelling for probes is largely one of **specific activity**, i.e. the measure of how radioactive the molecule is. For tracing purposes, a low specific activity will suffice, but for probes, a high specific activity is necessary. In probe preparation the radioactive label is usually the high-energy ß-emitter ^{32}P. Some common methods of labelling nucleic acid molecules are described below.

3.3.1 End labelling

In this technique the enzyme **polynucleotide kinase** is used to transfer the terminal phosphate group of ATP onto 5′-hydroxyl termini of nucleic acid molecules. If the ATP donor is radioactively labelled, this produces a labelled nucleic acid of relatively low specific activity, as only the termini of each molecule become radioactive (Fig. 3.2).

3.3.2 Nick translation

This method relies on the ability of the enzyme **DNA polymerase I** (see section 4.2.2) to translate (move along the DNA) a nick created in the

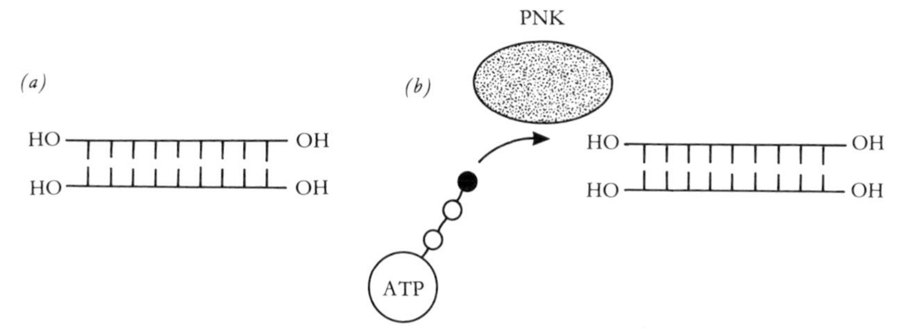

Fig. 3.2. End labelling DNA using polynucleotide kinase (PNK). (*a*) DNA is dephosphorylated using phosphatase, to generate 5′-OH groups. (*b*) The terminal phosphate of [γ-^{32}P]ATP (solid circle) is then transferred to the 5′ terminus by PNK. The reaction can also occur as an exchange reaction with 5′-phosphate termini.

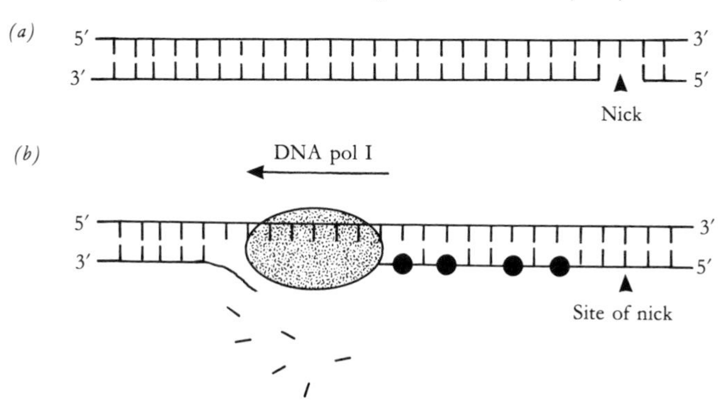

Fig. 3.3. Labelling DNA by nick translation. (*a*) A single-strand nick is introduced into the phosphodiester backbone of a DNA fragment using DNase I. (*b*) DNA polymerase (pol) I then synthesises a copy of the template strand, degrading the non-template strand with its 5′→3′ exonuclease activity. If [α-^{32}P]dNTP is supplied this will be incorporated into the newly synthesised strand (filled circles).

phosphodiester backbone of the DNA double helix. Nicks may occur naturally, or may be caused by a low concentration of the nuclease **DNase I** in the reaction mixture. DNA polymerase I catalyses a strand replacement reaction which incorporates new dNTPs into the DNA chain. If one of the dNTPs supplied is radioactive, the result is a highly labelled DNA molecule (Fig. 3.3).

3.3.3 Labelling by primer extension

This term refers to a technique which uses random **oligonucleotides** (usually hexadeoxyribonucleotide molecules – sequences of six deoxynuc-

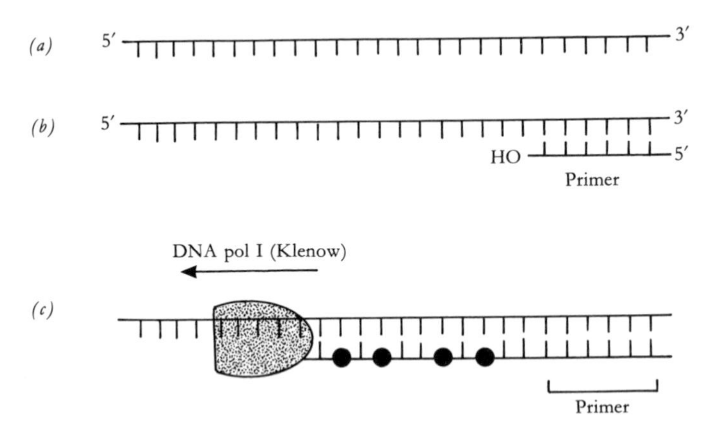

Fig. 3.4. Labelling DNA by primer extension (oligolabelling). (*a*) DNA is denatured to give single-stranded molecules. (*b*) An oligonucleotide primer is then added to give a short double-stranded region with a free 3′-OH group. (*c*) The Klenow fragment of DNA polymerase (pol) I can then synthesise a copy of the template strand from the primer, incorporating [α-^{32}P]dNTP (filled circles) to give a highly labelled molecule.

leotides) to prime synthesis of a DNA strand by DNA polymerase. The DNA to be labelled is denatured by heating, and the oligonucleotide primers annealed to the single stranded DNAs. The **Klenow fragment** of DNA polymerase (see section 4.2.2) can then synthesise a copy of the template, primed from the 3′-hydroxyl group of the oligonucleotide. If a labelled dNTP is incorporated, DNA of very high specific activity is produced (Fig. 3.4).

In a radiolabelling reaction it is often desirable to separate the labelled DNA from the unincorporated nucleotides present in the reaction mixture. A simple way of doing this is to carry out a small-scale gel filtration step using a suitable medium. The whole process can be carried out in a Pasteur pipette, with the labelled DNA coming off the column first, followed by the free nucleotides. Fractions can be collected and monitored for radioactivity, and the data used to calculate total activity of the DNA, specific activity, and percentage incorporation of the isotope.

3.4 Nucleic acid hybridisation

The complementary nature of base-pairing in DNA is one of its most useful features, and can be used to advantage in a variety of applications. If a DNA duplex is denatured by heating the solution until the strands separate, the

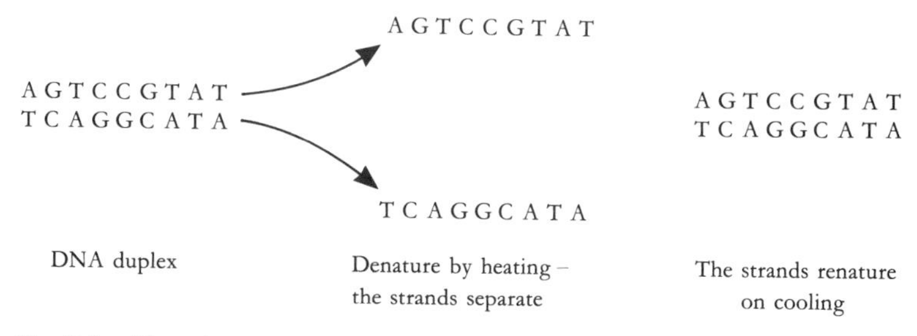

Fig. 3.5. The principle of nucleic acid hybridisation.

complementary strands will renature on cooling (Fig. 3.5). This feature can be used to provide information about the sequence complexity of the DNA in question, since sequences that are present as multiple copies in the genome will renature faster than sequences that are present as single copies only. By performing this type of analysis, eukaryotic DNA can be shown to be composed of four different abundance classes. Firstly, some DNA will form duplex structures almost instantly, because the denatured strands have regions such as **inverted repeats** or **palindromes**, which fold back on each other to give a hairpin loop structure. This class is commonly known as **foldback DNA**. Second fastest to re-anneal are **highly repetitive** sequences, which occur many times in the genome. Following these are **moderately repetitive** sequences, and finally there are the **unique** or **single copy** sequences, which rarely re-anneal under the conditions used for this type of analysis.

In addition to providing information about sequence complexity, nucleic acid hybridisation can be used as an extremely sensitive detection method, capable of picking out specific DNA sequences from complex mixtures. Usually a single pure sequence is labelled with ^{32}P and used as a probe. The probe is denatured before use so that the strands are free to base-pair with their complements. The DNA to be probed is also denatured, and is usually fixed to a supporting membrane made from nitrocellulose or nylon. Hybridisation is carried out in a sealed plastic bag at 65–68 °C for several hours to allow the duplexes to form. The excess probe is then washed off and the degree of hybridisation can be monitored by counting the sample in a scintillation spectrometer or preparing an **autoradiogram**, where the sample is exposed to X-ray film. Some specific applications of nucleic acid hybridisation will be discussed in Chapter 7.

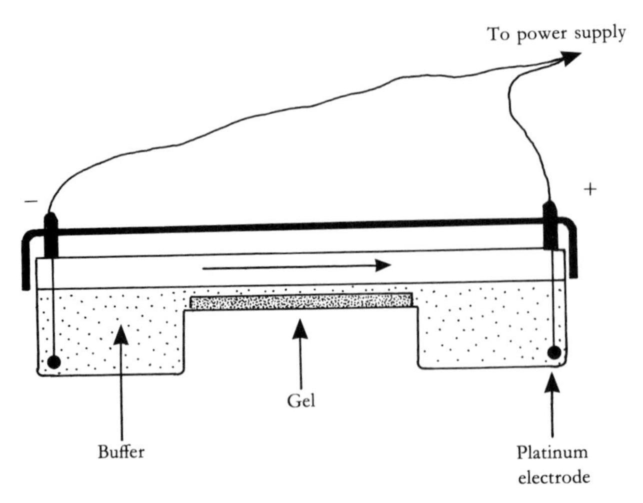

Fig. 3.6. A typical system used for agarose gel electrophoresis. The gel is just covered with buffer, therefore the technique is sometimes called **submerged agarose gel electrophoresis** (SAGE). Nucleic acid samples placed in the gel will migrate towards the positive electrode as indicated by the horizontal arrow.

3.5 Gel electrophoresis

The technique of gel electrophoresis is vital to the genetic engineer, as it represents the main way by which nucleic acid fragments may be visualised directly. The method relies on the fact that nucleic acids are **polyanionic** at neutral pH, i.e. they carry multiple negative charges due to the phosphate groups on the phosphodiester backbone of the nucleic acid strands. This means that the molecules will migrate towards the positive electrode when placed in an electric field. The technique is carried out using a gel matrix, which separates the nucleic acid molecules according to size. A typical nucleic acid electrophoresis setup is shown in Fig. 3.6.

The type of matrix used for electrophoresis has important consequences for the degree of separation achieved, which is dependent on the porosity of the matrix. Two gel types are commonly used, these being **agarose** and **polyacrylamide**. Agarose is extracted from seaweed, and can be purchased as a dry powder which is melted in buffer at an appropriate concentration, normally in the range 0.3–2.0% (w/v). On cooling, the agarose sets to form the gel. Agarose gels are usually run in the apparatus shown in Fig. 3.6. Polyacrylamide gels are sometimes used to separate small nucleic acid molecules in applications such as DNA sequencing (see section 3.6), as the pore size is smaller than that achieved with agarose. The useful separation ranges of agarose and polyacrylamide gels are shown in Table 3.1.

Table 3.1. *Separation characteristics for agarose and polyacrylamide gels*

Gel type	Separation range (base-pairs)
0.3% agarose	50 000 to 1000
0.7% agarose	20 000 to 300
1.4% agarose	6000 to 300
4% acrylamide	1000 to 100
10% acrylamide	500 to 25
20% acrylamide	50 to 1

Source: From Schleif and Wensink (1981), *Practical Methods in Molecular Biology,* Springer-Verlag, New York. Reproduced with permission.

Electrophoresis is carried out by placing the nucleic acid samples in the gel and applying a potential difference across it. This is maintained until a marker dye (usually bromophenol blue, added to the sample prior to loading) reaches the end of the gel. The nucleic acids in the gel are usually visualised by staining with the intercalating dye ethidium bromide and examining under u.v. light. Nucleic acids show up as orange bands, which can be photographed to provide a record (Fig. 3.7). The data can be used to estimate the sizes of unknown fragments by construction of a calibration curve using standards of known size, as migration is inversely proportional to the $\log_{10}$ of the number of base-pairs. This is particularly useful in the technique of **restriction mapping** (see section 4.1.3).

In addition to its use in the analysis of nucleic acids, polyacrylamide gel electrophoresis (PAGE) is used extensively for the analysis of proteins. The methodology is different from that used for nucleic acids, but the basic principles are similar.

3.6 DNA sequencing

The ability to determine the sequence of bases in DNA is a central part of modern molecular biology, and provides what might be considered as the ultimate structural information. Rapid methods for sequence analysis were developed in the late 1970s, and the technique is now used in laboratories worldwide.

There are two main methods for sequencing DNA. In one method, developed by Allan Maxam and Walter Gilbert, chemicals are used to cleave the DNA at certain positions, generating a set of fragments that differ by one nucleotide. The same result is achieved in a different way in the

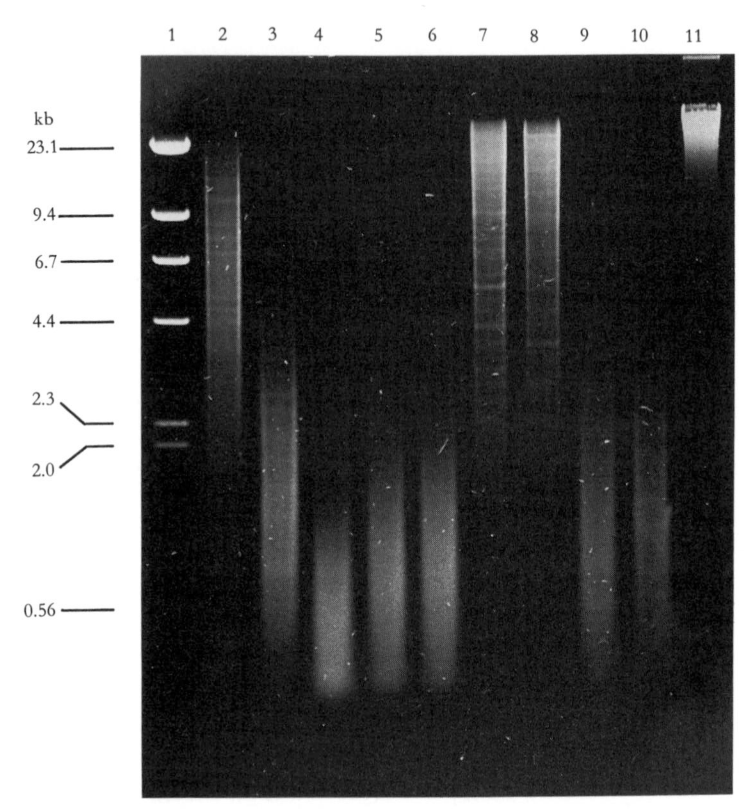

Fig. 3.7. Black-and-white photograph of an agarose gel, stained with ethidium bromide, under u.v. irradiation. The DNA samples show up as orange smears or as orange bands on a purple background. Individual bands (lane 1) indicate discrete fragments of DNA – in this case, the fragments are of phage λ DNA cut with the restriction enzyme *Hin*dIII. The sizes of the fragments (in kb) are indicated. The remaining lanes contain samples of DNA from an alga, cut with various restriction enzymes. Because of the heterogeneous nature of these samples, the fragments merge into one another and show up as a smear on the gel. Samples that have migrated farthest (lanes 3,4,5,6,9 and 10) are made up of smaller fragments than those that have remained near the top of the gel (lanes 2,7,8 and 11). Photograph courtesy of Dr N. Urwin.

second method, developed by Fred Sanger and Alan Coulson, which involves enzymatic synthesis of DNA strands that terminate in a modified nucleotide. Analysis of fragments is similar for both methods and involves gel electrophoresis and autoradiography. The enzymatic method has now largely replaced the chemical method as the technique of choice, although there are some situations where chemical sequencing can provide data more easily than the enzymatic method.

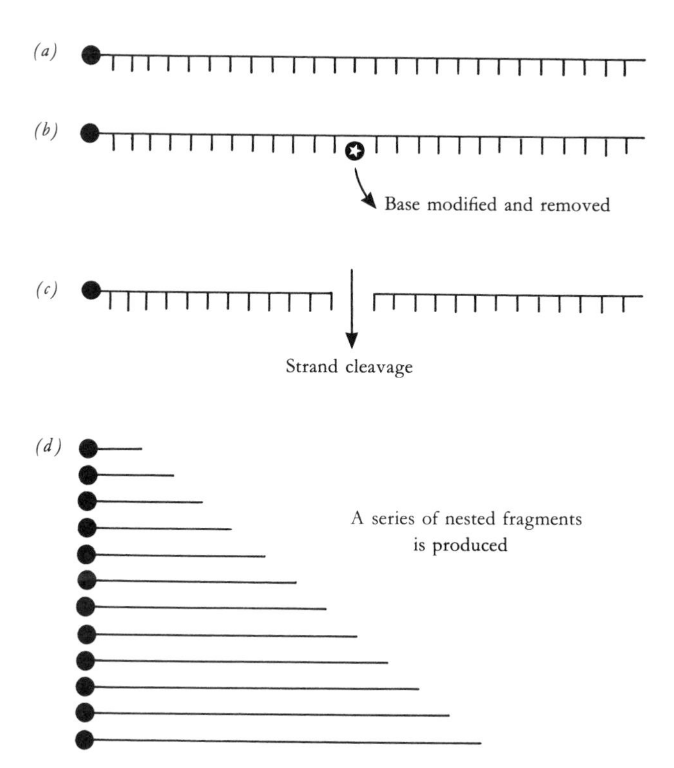

Fig. 3.8. DNA sequencing using the chemical (Maxam–Gilbert) method. (*a*) Radiolabelled single-stranded DNA is produced. (*b*) The bases in the DNA are chemically modified and removed, with, on average, one base being affected per molecule. (*c*) The phosphodiester backbone is then cleaved using piperidine. (*d*) The process produces a set of fragments differing in length by one nucleotide, labelled at their 5′ termini.

3.6.1 Maxam–Gilbert (chemical) sequencing

A defined fragment of DNA is required as the starting material. This need not be cloned in a plasmid vector, so the technique is applicable to any DNA fragment. The DNA is radiolabelled with ^{32}P at the 5′ ends of each strand, and the strands denatured, separated and purified to give a population of labelled strands for the sequencing reactions (Fig. 3.8). The next step is a chemical modification of the bases in the DNA strand. This is done in a series of four or five reactions with different specificities, and the reaction conditions are chosen so that, on average, only one modification will be introduced into each copy of the DNA molecule. The modified bases are then removed from their sugar groups and the strands cleaved at these positions using the chemical piperidine. The theory is that, given the large number of molecules and the different reactions, this process will produce a set of fragments which terminate at different bases and differ in length by one nucleotide. This is known as a set of **nested** fragments.

3.6.2 Sanger–Coulson (dideoxy or enzymatic) sequencing

Although the end result is similar to that attained by the chemical method, the Sanger–Coulson procedure is totally different from that of Maxam and Gilbert. In this case a copy of the DNA to be sequenced is made by the Klenow fragment of DNA polymerase (see section 4.2.2). The template for this reaction is single stranded DNA, and a primer must be used to provide the 3′ terminus for DNA polymerase to begin synthesising the copy (Fig. 3.9). The production of nested fragments is achieved by the incorporation of a modified dNTP in each reaction. These dNTPs lack a hydroxyl group at the 3′ position of deoxyribose, which is necessary for chain elongation to proceed. Such modified dNTPs are known as **dideoxynucleoside triphosphates** (ddNTPs). The four ddNTPs (A,G,T and C forms) are included in a series of four reactions, each of which contains the four normal dNTPs. The concentration of the dideoxy form is such that it will be incorporated into the growing DNA chain infrequently. Each reaction therefore produces a series of fragments terminating at a specific nucleotide, and the four reactions together provide a set of nested fragments. The DNA chain is labelled by including a radioactive dNTP in the reaction mixture. This is usually [α-^{35}S]dATP, which enables more sequence to be read from a single gel than the ^{32}P-labelled dNTPs that were used previously.

The generation of fragments for dideoxy sequencing is more complicated than for chemical sequencing, and usually involves subcloning into different vectors. One method is to clone the DNA into the bacteriophage M13 (see section 5.3.3), which produces single-stranded DNA during infection. This provides a suitable substrate for the sequencing reactions.

3.6.3 Electrophoresis and reading of sequences

Separation of the DNA fragments created in sequencing reactions is achieved by polyacrylamide gel electrophoresis. The gels usually contain 6–20% polyacrylamide and 7 M urea, which acts as a denaturant to reduce the effects of DNA secondary structure. This is important because fragments

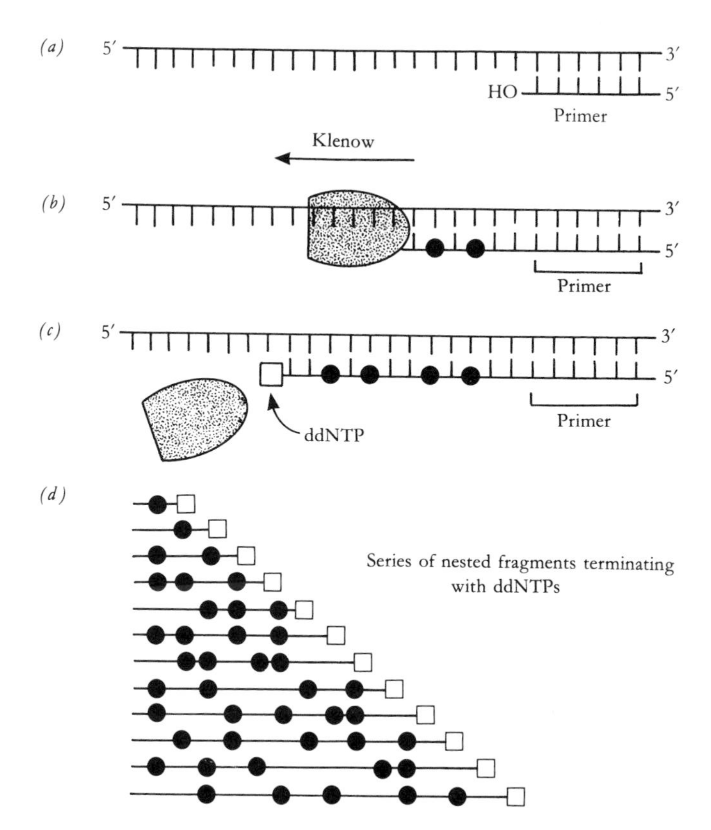

Fig. 3.9. DNA sequencing using the dideoxy chain termination (Sanger–Coulson) method. (*a*) A primer is annealed to a single-stranded template and (*b*) the Klenow fragment of DNA polymerase I used to synthesise a copy of the DNA. A radiolabelled dNTP (often [α-^{35}S]dNTP, filled circles) is incorporated into the DNA. (*c*) Chain termination occurs when a dideoxynucleoside triphosphate (ddNTP) is incorporated. (*d*) A series of four reactions, each containing one ddNTP in addition to the four dNTPs required for chain elongation, generates a set of radiolabelled nested fragments.

that differ in length by only one base are being separated. The gels are very thin (0.5 mm or less) and are run at high power settings, which causes them to heat up to 60–70 °C. This also helps to maintain denaturing conditions. Sometimes two lots of samples are loaded onto the same gel at different times to maximise the amount of sequence information obtained.

When the gel has been run, it is removed from the apparatus and may be dried onto a paper sheet to facilitate handling. It is then exposed to X-ray film. The emissions from the radioactive label sensitise the silver grains, which turn black when the film is developed and fixed. The result is known as an autoradiogram (Fig.3.10(*a*)). Reading the autoradiogram is straight-

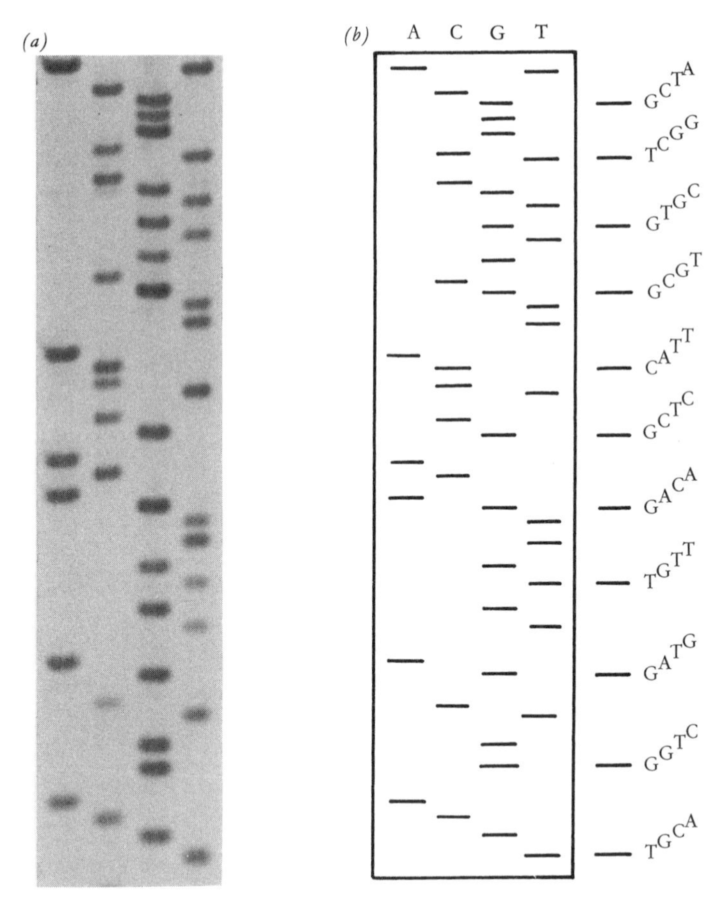

Fig. 3.10. Reading a DNA sequence. (*a*) An autoradiogram of part of a sequencing gel, and (*b*) a tracing of the autoradiogram. Each lane corresponds to a reaction containing one of the four ddNTPs. The sequence is read from the bottom of the gel, each successive fragment being one nucleotide longer than the preceding one. Photograph courtesy of Dr N. Urwin.

forward – the sequence is read from the smallest fragment upwards, as shown in Fig. 3.10(*b*). Using this method sequences of up to several hundred bases may be read from single gels. The sequence data are then compiled and studied using a computer, which can perform analyses such as translation into amino acid sequences and identification of restriction sites, regions of sequence homology and other structural motifs such as promoters and control regions.

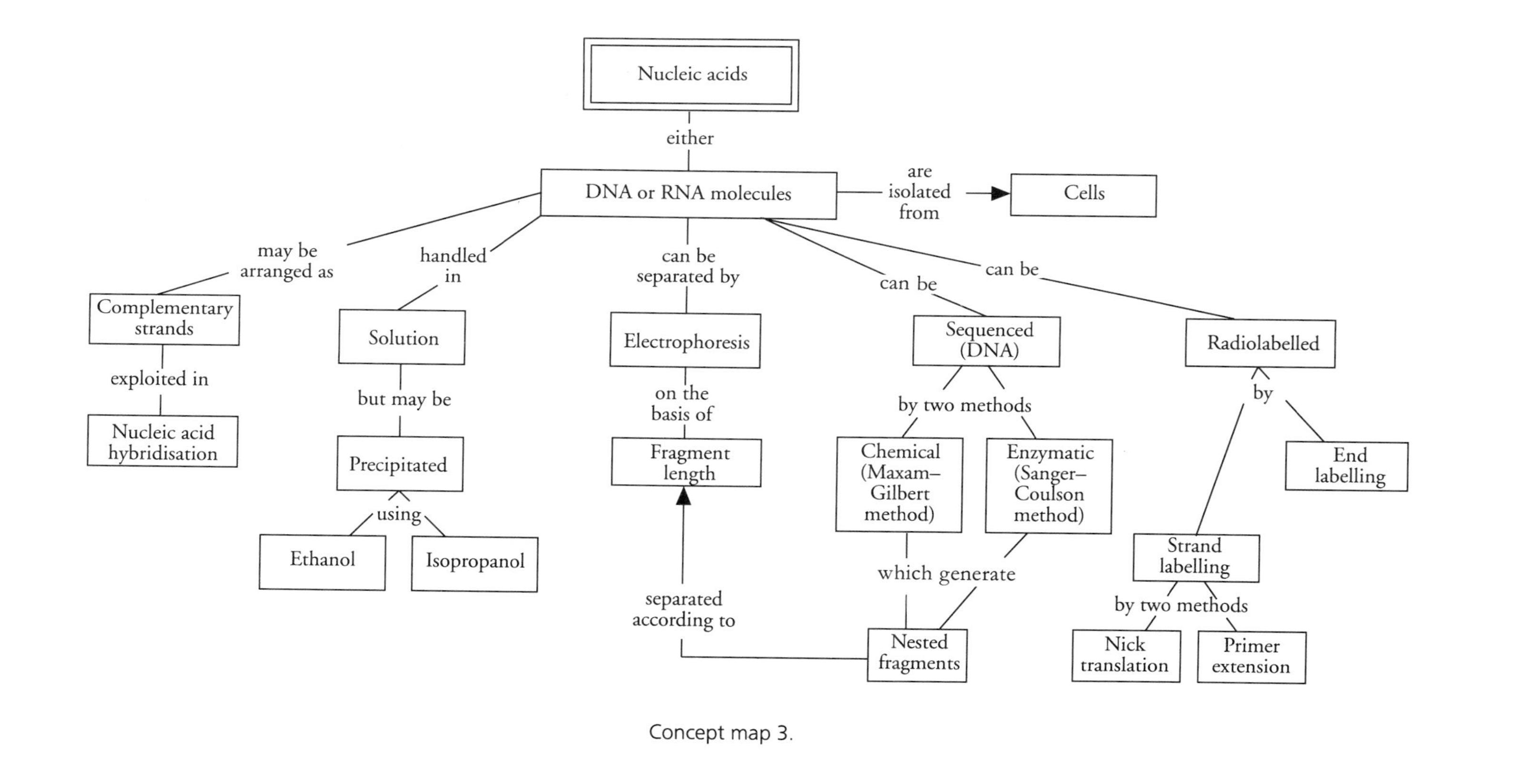

Concept map 3.

4

The tools of the trade

The genetic engineer needs to be able to cut and join DNA from different sources. In addition, certain modifications may have to be carried out to the DNA during the various steps required to produce, clone and identify recombinant DNA molecules. The tools that enable these manipulations to be performed are **enzymes**, which are purified from a wide range of organisms and can be bought from various suppliers. In this chapter I examine some of the important classes of enzymes that make up the genetic engineer's toolkit.

4.1 Restriction enzymes – cutting DNA

The **restriction enzymes**, which cut DNA at defined sites, represent one of the most important groups of enzymes for the manipulation of DNA. These enzymes are found in bacterial cells, where they function as part of a protective mechanism called the **restriction–modification** system. In this system the restriction enzyme hydrolyses any exogenous DNA that appears in the cell. To prevent the enzyme acting on the host cell DNA, the modification enzyme of the system (a methylase) modifies the host DNA by methylation of particular bases in the recognition sequence, which prevents the restriction enzyme from cutting the DNA.

Restriction enzymes are of three types (I, II or III). Most of the enzymes used today are type II enzymes, which have the simplest mode of action. These enzymes are **nucleases** (see section 4.2.1), and as they cut at an internal position in a DNA strand (as opposed to beginning degradation at one end) they are known as **endonucleases**. Thus the correct designation of such enzymes is that they are type II restriction endonucleases, although they are often simply called restriction enzymes. In essence they may be thought of as molecular scissors.

4.1.1 Type II restriction endonucleases

Restriction enzyme nomenclature is based on a number of conventions. The generic and specific names of the organism in which the enzyme is found are used to provide the first part of the designation, which comprises the first letter of the generic name and the first two letters of the specific name. Thus an enzyme from a strain of *Escherichia coli* is termed *Eco*, one from *Bacillus amyloliquefaciens* is *Bam*, and so on. Further descriptors may be added, depending on the bacterial strain involved and on the presence or absence of extrachromosomal elements. Two widely used enzymes from the bacteria mentioned above are *Eco*RI and *Bam*HI.

The value of restriction endonucleases lies in their specificity. Each particular enzyme recognises a specific sequence of bases in the DNA, the most common recognition sequences being four, five or six base-pairs in length. Thus, given that there are four bases in the DNA, and assuming a random distribution of bases, the expected frequency of any particular sequence can be calculated as 4^n, where n is the length of the recognition sequence. This predicts that tetranucleotide sites will occur every 256 base-pairs, pentanucleotide sites every 1024 base-pairs, and hexanucleotide sites every 4096 base-pairs. There is, as you might expect, considerable variation from these values, but generally the fragment lengths produced will lie around the calculated value. Thus an enzyme recognising a tetranucleotide sequence (sometimes called a 'four-cutter') will produce shorter DNA fragments than a six-cutter. Some of the most commonly used restriction enzymes are listed in Table 4.1, with their recognition sequences and cutting sites.

4.1.2 Use of restriction endonucleases

Restriction enzymes are very simple to use – an appropriate amount of enzyme is added to the target DNA in a buffer solution, and the reaction is incubated at 37 °C. Enzyme activity is expressed in units, with one unit being the amount of enzyme that will cleave one microgram of DNA in one

Table 4.1. *Recognition sequences and cutting sites for some restriction endonucleases*

Enzyme	Recognition sequence	Cutting sites	Ends
*Bam*HI	5′-GGATCC-3′	G↓G A T C C C C T A G↑G	5′
*Eco*RI	5′-GAATTC-3′	G↓A A T T C C T T A A↑G	5′
*Hae*III	5′-GGCC-3′	G G↓C C C C↑G G	Blunt
*Hpa*I	5′-GTTAAC-3′	G T T↓A A C C A A↑T T G	Blunt
*Pst*I	5′-CTGCAG-3′	C T G C A↓G G↑A C G T C	3′
*Sau*3A	5′-GATC-3′	↓G A T C C T A G↑	5′
*Sma*I	5′-CCCGGG-3′	C C C↓G G G G G G↑C C C	Blunt
*Sst*I	5′-GAGCTC-3′	G A G C T↓C C↑T C G A G	3′
*Xma*I	5′-CCCGGG-3′	C↓C C G G G G G G C C↑C	5′

Note: The recognition sequences are given in single-strand form, written 5′→3′. Cutting sites are given in double-stranded form to illustrate the type of ends produced by a particular enzyme; 5′ and 3′ refer to 5′ and 3′-protruding termini, respectively.

hour at 37 °C. Although most experiments require complete digestion of the target DNA, there are some cases where various combinations of enzyme concentration and incubation time may be used to achieve only partial digestion (see section 6.3.2).

The type of DNA fragment that a particular enzyme produces depends on the recognition sequence and on the location of the cutting site within this sequence. As we have already seen, fragment length is dependent on the frequency of occurrence of the recognition sequence. The actual cutting site of the enzyme will determine the type of ends that the cut fragment has, which is important with regard to further manipulation of the DNA. Three types of fragment may be produced, these being (i) blunt or flush-ended fragments, (ii) fragments with protruding 3′ ends, and (iii) fragments with protruding 5′ ends. An example of each type is shown in Fig. 4.1.

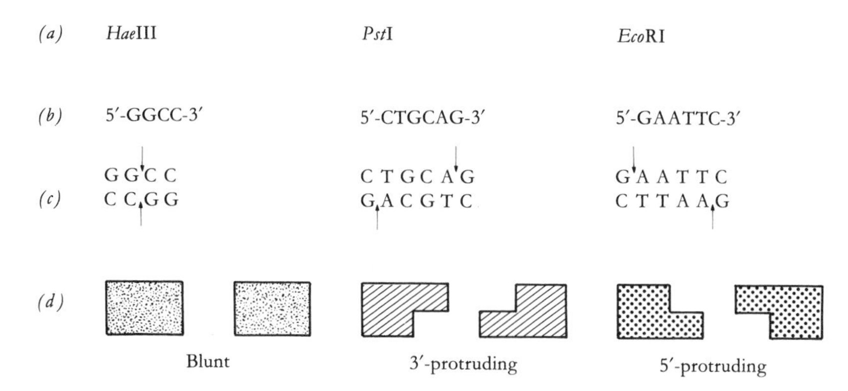

Fig. 4.1. Types of ends generated by different restriction enzymes. The enzymes are listed in (*a*), with their recognition sequences and cutting sites shown in (*b*) and (*c*), respectively. (*d*) A schematic representation of the types of ends generated.

Enzymes such as *Pst*I and *Eco*RI generate DNA fragments with cohesive or 'sticky' ends, as the protruding sequences can base pair with complementary sequences generated by the same enzyme. Thus, by cutting two different DNA samples with the same enzyme and mixing the fragments together, recombinant DNA can be produced, as shown in Fig. 4.2. This is one of the most useful applications of restriction enzymes, and is a vital part of many manipulations in genetic engineering.

4.1.3 Restriction mapping

Most pieces of DNA will have recognition sites for various restriction enzymes, and it is often beneficial to know the relative locations of some of these sites. The technique used to obtain this information is known as **restriction mapping**. This involves cutting a DNA fragment with a selection of restriction enzymes, singly and in various combinations. The fragments produced are run on an agarose gel and their sizes determined. From the data obtained, the relative locations of the cutting sites can be worked out. A fairly simple example can be used to illustrate the technique, as outlined below.

Let us say that we wish to map the cutting sites for the restriction enzymes *Bam*HI, *Eco*RI and *Pst*I, and that the DNA fragment of interest is 15 kb in length. Various digestions are carried out, and the fragments arising from these are analysed and their sizes determined. The results obtained are shown in Table 4.2. As each of the single enzyme reactions produces two DNA fragments, we can conclude that the DNA has a single

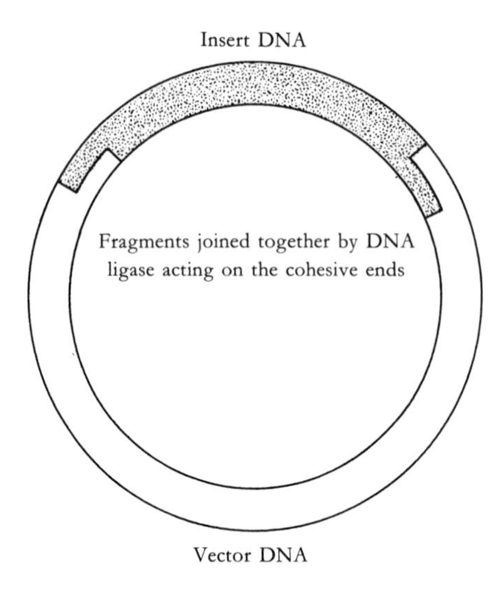

Fig. 4.2. Generation of recombinant DNA. DNA fragments from different sources can be joined together if they have cohesive ('sticky') ends, as produced by many restriction enzymes. On annealing the complementary regions, the phosphodiester backbone is sealed using DNA ligase.

Table 4.2. *Digestion of a 15 kb DNA fragment with three restriction enzymes*

*Bam*HI	*Eco*RI	*Pst*I	*Bam*HI + *Eco*RI	*Bam*HI + *Pst*I	*Eco*RI + *Pst*I	*Bam*HI + *Eco*RI + *Pst*I
14	12	8	11	8	7	6
1	3	7	3	6	5	5
			1	1	3	3
						1

Note: Data shown are lengths (in kb) of fragments that are produced on digestion of a 15 kb DNA fragment with the enzymes *Bam*HI, *Eco*RI and *Pst*I. Single, double and triple digests were carried out as indicated.

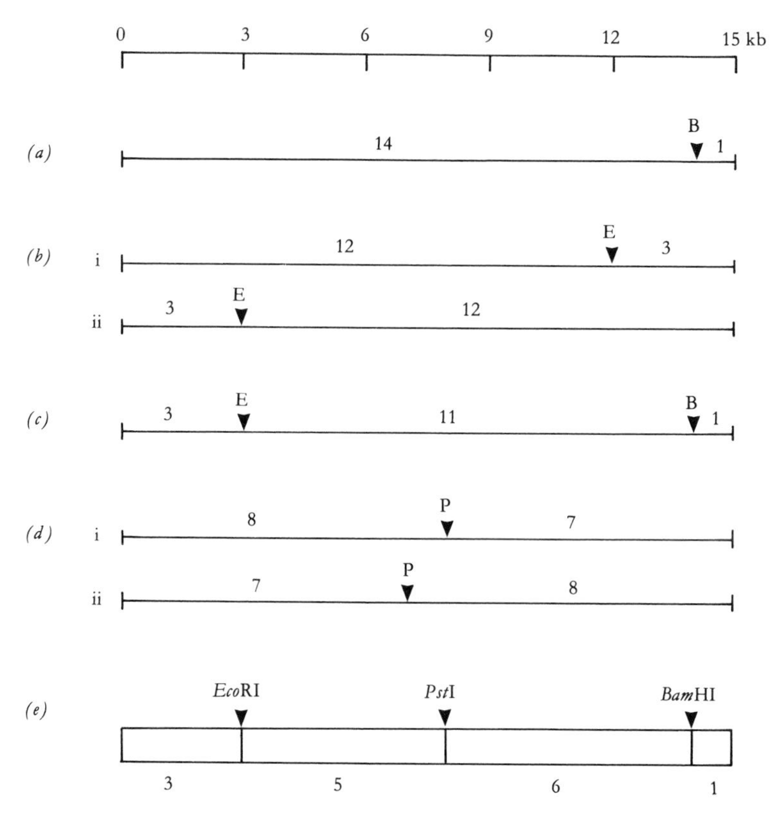

Fig. 4.3. Restriction mapping. (*a*) The 15 kb fragment yields two fragments of 14 and 1 kb when cut with *Bam*HI. (*b*) The *Eco*RI fragments of 12 and 3 kb can be orientated in two ways with respect to the *Bam*HI site. The *Bam*HI/*Eco*RI double digest gives fragments of 11, 3 and 1 kb, and therefore the relative positions of the *Bam*HI and *Eco*RI sites are as shown in (*c*). Similar reasoning with the orientation of the 8 and 7 kb *Pst*I fragments (*d*) gives the final map (*e*).

cutting site for each enzyme. The double digests enable a map to be drawn up, and the triple digest confirms this. Construction of the map is outlined in Fig. 4.3.

4.2 DNA modifying enzymes

Restriction enzymes (described above) and DNA ligase (section 4.3) provide the cutting and joining functions that are essential for the production of recombinant DNA molecules. Other enzymes used in genetic engineering may be loosely termed DNA **modifying** enzymes, with the term used here to include degradation, synthesis and alteration of DNA. Some of the most commonly used enzymes are described below.

4.2.1 Nucleases

Nuclease enzymes degrade nucleic acids by breaking the phosphodiester bond that holds the nucleotides together. Restriction enzymes are good examples of **endonucleases**, which cut within a DNA strand. A second group of nucleases, which degrade DNA from the termini of the molecule, are known as **exonucleases**.

Apart from restriction enzymes, there are four nucleases that are often used in genetic engineering. These are **Bal 31** and **exonuclease III** (exonucleases); and **deoxyribonuclease I** (DNase I) and **S_1-nuclease** (endonucleases). These enzymes differ in their precise mode of action, and provide the genetic engineer with a variety of strategies for attacking DNA. Their features are summarised in Fig. 4.4.

In addition to DNA-specific nucleases, there are **ribonucleases**, which act on RNA. These may be required for many of the stages in the preparation and analysis of recombinants, but as they are not directly involved in the construction of recombinant DNA molecules, they will not be described in detail.

4.2.2 Polymerases

Polymerase enzymes synthesise copies of nucleic acid molecules, and are used in many genetic engineering procedures. When describing a polymerase enzyme, the terms 'DNA-dependent' or 'RNA-dependent' may be used to indicate the type of nucleic acid template that the enzyme uses. Thus a DNA-dependent DNA polymerase copies DNA into DNA, an RNA-dependent DNA polymerase copies RNA into DNA, and a DNA-dependent RNA polymerase transcribes DNA into RNA. These enzymes synthesise nucleic acids by joining together nucleotides whose bases are complementary to the template strand bases. The synthesis proceeds in a $5' \rightarrow 3'$ direction, as each subsequent nucleotide addition requires a free 3'-OH group for the formation of the phosphodiester bond. This requirement also means that a short double-stranded region with an exposed 3'-OH (a primer) is necessary for synthesis to begin.

The enzyme **DNA polymerase I** has, in addition to its polymerase function, $5' \rightarrow 3'$ and $3' \rightarrow 5'$ exonuclease activities. The enzyme catalyses a

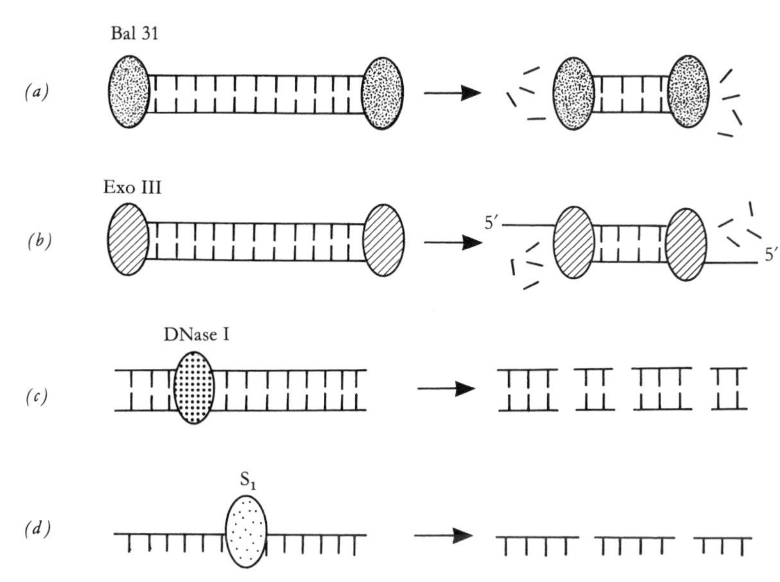

Fig. 4.4. Mode of action of various nucleases. (*a*) Nuclease Bal 31 is a complex enzyme. Its primary activity is a fast-acting 3′ exonuclease, which is coupled with a slow-acting endonuclease. When Bal 31 is present at a high concentration these activities effectively shorten DNA molecules from both termini. (*b*) Exonuclease III is a 3′ exonuclease that generates molecules with protruding 5′ termini. (*c*) DNase I cuts either single-stranded or double-stranded DNA at essentially random sites. (*d*) Nuclease S_1 is specific for single-stranded RNA or DNA. Modified from Brown (1990), *Gene Cloning*, Chapman and Hall; and Williams and Patient (1988), *Genetic Engineering*, IRL Press. Reproduced with permission.

strand replacement reaction, where the 5′→3′ exonuclease function degrades the non-template strand as the polymerase synthesises the new copy. A major use of this enzyme is in the nick translation procedure for radiolabelling DNA (outlined in section 3.3.2).

The 5′→3′ exonuclease function of DNA polymerase I can be removed by cleaving the enzyme to produce what is known as the **Klenow fragment**. This retains the polymerase and 3′→5′ exonuclease activities. The Klenow fragment is used where a single-stranded DNA molecule needs to be copied; because the 5′→3′ exonuclease function is lacking, the enzyme cannot degrade the non-template strand of dsDNA during synthesis of the new DNA. The 3′→5′ exonuclease activity is supressed under the conditions normally used for the reaction. Major uses for the Klenow fragment include radiolabelling by primed synthesis and DNA sequencing by the dideoxy method (see sections 3.3.3 and 3.6.2) in addition to the copying of single-stranded DNAs during the production of recombinants.

Reverse transcriptase (RTase) is an RNA-dependent DNA polymerase, and therefore produces a DNA strand from an RNA template. It has no associated exonuclease activity. The enzyme is used mainly for copying mRNA molecules in the preparation of cDNA (**complementary** or **copy** DNA) for cloning (see section 6.2.1), although it will also act on DNA templates.

4.2.3 Enzymes that modify the ends of DNA molecules

The enzymes **alkaline phosphatase**, **polynucleotide kinase** and **terminal transferase** act on the termini of DNA molecules, and provide important functions that are used in a variety of ways. The phosphatase and kinase enzymes, as their names suggest, are involved in the removal or addition of phosphate groups. Bacterial alkaline phosphatase (BAP: there is also a similar enzyme, calf intestinal alkaline phosphatase, CIP) removes phosphate groups from the 5′ ends of DNA, leaving a 5′-OH group. The enzyme is used to prevent unwanted ligation of DNA molecules, which can be a problem in certain cloning procedures. It is also used prior to the addition of radioactive phosphate to the 5′ ends of DNAs by polynucleotide kinase (see section 3.3.1).

Terminal transferase (terminal deoxynucleotidyl transferase) repeatedly adds nucleotides to any available 3′ terminus. Although it works best on protruding 3′ ends, conditions can be adjusted so that blunt-ended or 3′-recessed molecules may be utilised. The enzyme is mainly used to add homopolymer tails to DNA molecules prior to the construction of recombinants (see section 6.2.2).

4.3 DNA ligase – joining DNA molecules

DNA ligase is an important cellular enzyme, as its function is to repair broken phosphodiester bonds that may occur at random or as a consequence of DNA replication or recombination. In genetic engineering it is used to seal discontinuities in the sugar–phosphate chains that arise when recombinant DNA is made by joining DNA molecules from different sources. It can therefore be thought of as molecular glue, which is used to stick pieces of DNA together. This function is crucial to the success of many experiments, and DNA ligase is therefore a key enzyme in genetic engineering.

The enzyme used most often in experiments is T4 DNA ligase, which is

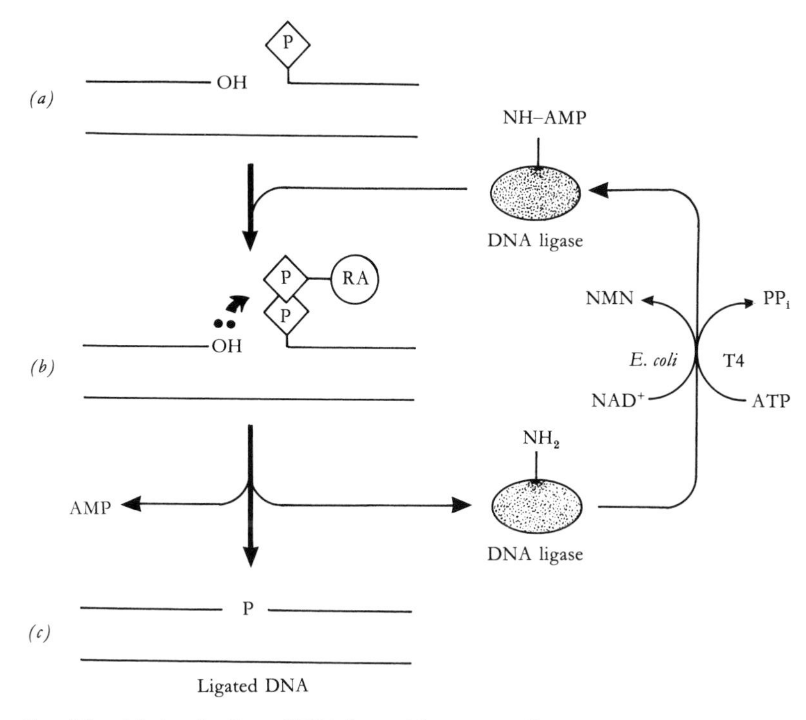

Fig. 4.5. Mode of action of DNA ligase. The enzyme ligates nucleotides *via* 3′-OH and $5'\text{-}PO_3^-$ termini that are adjacent due to a nick (*a*). DNA ligase is adenylated by NAD^+ (*E. coli*) or ATP (phage T4). The enzyme then adenylates the 5′ terminal phosphate at the nick (RA is ribose–adenine), which enables phosphodiester bond formation to proceed *via* nucleophilic attack (*b, c*). From Mathews and van Holde (1990), *Biochemistry*, Benjamin/Cummings. Reproduced with permission.

purified from *E. coli* cells infected with bacteriophage T4. Although the enzyme is most efficient when sealing gaps in fragments that are held together by cohesive ends, it will also join blunt-ended DNA molecules together under appropriate conditions. The enzyme works best at 37 °C, but is used at much lower temperatures (4–15 °C) to prevent thermal denaturation of the short base-paired regions that hold the cohesive ends of DNA molecules together. The mode of action of DNA ligase is outlined in Fig. 4.5.

The ability to cut, modify and join DNA molecules gives the genetic engineer the freedom to create recombinant DNA molecules. The technology involved is a test-tube technology, with no requirement for a living system. However, once a recombinant DNA fragment has been generated *in vitro*, it usually has to be amplified so that enough material is available for

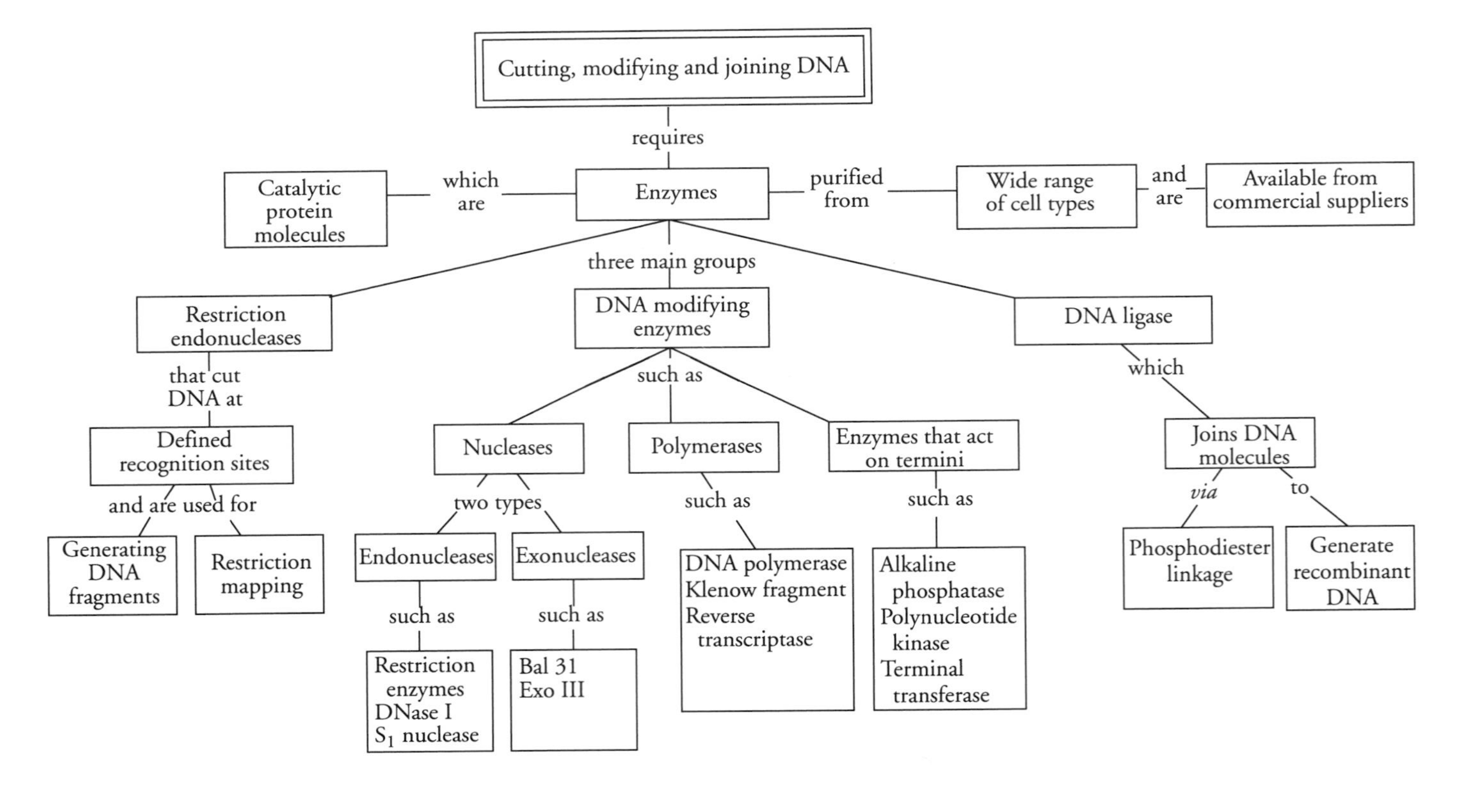

Concept map 4.

subsequent manipulation and analysis. Amplification requires a biological system, and we must therefore examine the types of living systems that can be used for the propagation of recombinant DNA molecules. These systems are described in the next chapter.

5

The biology of genetic engineering

Once recombinant DNA molecules have been constructed *in vitro*, the desired sequence can be isolated. Three things have to be done to achieve this: (i) the individual recombinant molecules have to be physically separated from each other, (ii) the recombinant sequences have to be amplified to provide enough material for further analysis, and (iii) the specific fragment of interest has to be selected by some sort of sequence-dependent method. In this chapter I consider the first two of these requirements, which in essence represent the techniques of **gene cloning**. Methods for selecting specific sequences are described in Chapter 7.

In some experiments hundreds of thousands of different DNA fragments may be produced, and the isolation of a particular sequence would seem to be an almost impossible task. It is a bit like looking for the proverbial needle in a haystack – with the added complication that the needle is made of the same material as the haystack! Fortunately the methods available provide a relatively simple way to achieve the desired outcome, i.e. the physical isolation of DNA sequences and their amplification and selection.

In this chapter I look at the biology of gene cloning, which involves the use of a suitable carrier molecule or **vector**, and requires a suitable living system or **host** in which the vector can be propagated. Types of host cell will be described first, followed by vector systems and methods for getting DNA into cells.

5.1 Host cell types

The type of host cell used for a particular application will depend mainly on the purpose of the cloning experiment. If the aim is to isolate a gene for

Table 5.1. *Types of host cell used for genetic engineering*

Major group	Prokaryotic/ eukaryotic	Type	Examples
Bacteria	Prokaryotic	Gram −ve	*Escherichia coli*
		Gram +ve	*Bacillus subtilis*
			Streptomyces spp.
Fungi	Eukaryotic	Microbial	*Saccharomyces cerevisiae*
		Filamentous	*Aspergillus nidulans*
Plants	Eukaryotic	Protoplasts	Various types
		Intact cells	Various types
		Whole organism	Various types
Animals	Eukaryotic	Insect cells	*Drosophila melanogaster*
		Mammalian cells	Various types
		Oocytes	Various types
		Whole organism	Various types

Note: Plant and animal cells may be subjected to manipulation either in tissue culture or as cells in the whole organism. In some cases cells in culture may be hybrids formed by fusing cells of different species.

structural analysis, the requirements may call for a simple system that is easy to use. If the aim is to express the genetic information in a higher eukaryote such as a plant, a more specific system will be required. These two aims are not necessarily mutually exclusive; often a simple **primary** host is used to isolate a sequence that is then introduced into a more complex system for expression. The main types of host cell are shown in Table 5.1, and are described below.

5.1.1 Prokaryotic hosts

An ideal host cell should be easy to handle and propagate, should be available as a wide variety of genetically defined strains, and should accept a range of vectors. The bacterium *Escherichia coli* fulfils these requirements, and is used in many cloning protocols. *E. coli* has been studied in great detail, and many different strains were isolated by microbial geneticists as they investigated the genetic mechanisms of this prokaryotic organism. Such studies provided the essential background information on which genetic engineering is based.

E. coli is a Gram-negative bacterium with a single chromosome packed into a compact structure known as the **nucleoid**. The genome size is some 4×10^6 base-pairs. The processes of gene expression (transcription and translation) are coupled, with the newly synthesised mRNA being immediately available for translation. There is no post-transcriptional modification of the primary transcript as is commonly found in eukaryotic cells. *E. coli* can therefore be considered as one of the simplest host cells. Much of the gene cloning that is carried out routinely in laboratories involves the use of *E. coli* hosts, with many genetically different strains available for specific applications.

In addition to *E. coli*, other bacteria may be used as hosts for gene cloning experiments, with examples including species of *Bacillus*, *Pseudomonas* and *Streptomyces*. There are, however, certain drawbacks with most of these. Often there are few suitable vectors available for use in such cells, and getting recombinant DNA into the cell can cause problems. This is particularly troublesome when primary cloning experiments are envisaged, i.e. direct introduction of ligated recombinant DNA into the host cell. It is often more sensible to perform the initial cloning in *E. coli*, isolate the required sequence, and then introduce the purified DNA into the target host. Many of the drawbacks can be overcome by using this approach, particularly when vectors that can function in the target host and in *E. coli* (**shuttle vectors**) are used. Use of bacteria other than *E. coli* will not be discussed further in this book; details may be found in some of the texts mentioned in the section Suggestions for further reading.

5.1.2 Eukaryotic hosts

One disadvantage of using an organism such as *E. coli* as a host for cloning is that it is a prokaryote, and therefore lacks the membrane-bound nucleus (and other organelles) found in eukaryotic cells. This means that certain eukaryotic genes may not function in *E. coli* as they would in their normal environment, which can hamper their isolation by selection mechanisms that depend on gene expression. Also, if the production of a eukaryotic protein is the desired outcome of a cloning experiment, it may not be easy to ensure that a prokaryotic host produces a fully functional protein.

Eukaryotic cells range from microbes, such as yeast and algae, to cells from complex multicellular organisms, such as ourselves. The microbial cells have many of the characteristics of bacteria with regard to ease of growth and availability of mutants. Higher eukaryotes present a different set of problems to the genetic engineer, many of which require specialised solutions. Often the aim of a cloning experiment in a higher plant or animal is to alter the genetic makeup of the organism, rather than to isolate a gene for further analysis or to produce large amounts of a particular protein.

The yeast *Saccharomyces cerevisiae* is the favoured eukaryotic microbe for genetic engineering. It has been used for centuries in the production of bread and beer, and has been studied extensively. The organism is amenable to classical genetic analysis, and a range of mutant cell types is available. In terms of genome complexity, *S. cerevisiae* has about 1.35×10^7 base pairs of DNA, some 3.5 times more than *E. coli*. Other fungi that may be used for gene cloning experiments include *Aspergillus nidulans* and *Neurospora crassa*.

Plant and animal cells may also be used as hosts for gene manipulation experiments. Unicellular algae such as *Chlamydomonas reinhardtii* have all the advantages of microorganisms plus the structural and functional organisation of plant cells, and their use in genetic manipulation will increase as they become more widely studied. Other plant cells (and animal cells) are usually grown as cell cultures, which are much easier to manipulate than cells in a whole organism. Some aspects of genetic engineering in plant and animal cells are discussed in sections 8.3 and 8.4.

5.2 Plasmid vectors for use in *E. coli*

There are certain essential features that vectors must possess. Ideally they should be fairly small DNA molecules, to facilitate isolation and handling. There must be an **origin of replication**, so that their DNA can be copied and thus maintained in the cell population as the host organism grows and divides. It is desirable to have some sort of **selectable marker** that will enable the vector to be detected, and the vector must also have at least one unique restriction endonuclease recognition site, to enable DNA to be inserted during the production of recombinants. Plasmids have these features, and are extensively used as vectors in cloning experiments. Some features of plasmid vectors are described below.

5.2.1 What are plasmids?

Many types of plasmid are found in nature, in bacteria and some yeasts. They are circular DNA molecules, relatively small when compared to the host cell chromosome, that are maintained mostly in an extrachromosomal state. Although plasmids are generally dispensable (i.e. not essential for cell growth and division), they often confer traits (such as antibiotic resistance) on the host organism, which can be a selective advantage under certain conditions. The antibiotic resistance genes encoded by plasmid DNA (pDNA) are often used in the construction of vectors for genetic engineering, as they provide a convenient means of selecting cells containing the plasmid.

Plasmids can be classified into two groups, termed **conjugative** and **non-conjugative**. Conjugative plasmids can mediate their own transfer between bacteria by the process of conjugation, which requires functions specified by the *tra* (transfer) and *mob* (mobilising) regions carried on the plasmid. Non-conjugative plasmids are not self-transmissible, but may be mobilised by a conjugation-proficient plasmid if their *mob* region is functional. A further classification is based on the number of copies of the plasmid found in the host cell, a feature known as the **copy number**. Low copy number plasmids tend to exhibit **stringent** control of DNA replication, with replication of the pDNA closely tied to host cell chromosomal DNA replication. High copy number plasmids are termed **relaxed** plasmids, with DNA replication not dependent on host cell chromosomal DNA replication. In general terms, conjugative plasmids are large, show stringent control of DNA replication, and are present at low copy numbers, whilst non-conjugative plasmids are small, show relaxed DNA replication and are present at high copy numbers. Some examples of plasmids are shown in Table 5.2.

5.2.2 Basic cloning plasmids

For genetic engineering, naturally occurring plasmids have been extensively modified to produce vectors that have the desired characteristics. In naming plasmids, p is used to designate plasmid, and this is usually followed by the initials of the worker(s) who isolated or constructed the plasmid. Numbers may be used to classify the particular isolate. One of the most extensively used plasmids is pBR322, which was developed by Francisco Bolivar and his colleagues. Construction of pBR322 involved a series of manipulations to get the right pieces of DNA together, with the final result containing DNA from three sources. The plasmid has all the features of a good vector, such as low molecular weight, antibiotic resistance genes, an origin of replication, and several single-cut restriction endonuclease recognition sites. A map of pBR322 is shown in Fig. 5.1.

There are several plasmids in the pBR series, each with slightly different

Table 5.2. *Properties of some naturally occuring plasmids*

Plasmid	Size (kb)	Conjugative	Copy number	Selectable markers
ColE1	7.0	No	10–15	$E1^{imm}$
RSF1030	9.3	No	20–40	Ap^r
CloDF13	10.0	No	10	$DF13^{imm}$
pSC101	9.7	No	1–2	Tc^r
RK6	42	Yes	10–40	Ap^r, Sm^r
F	103	Yes	1–2	—
R1	108	Yes	1–2	Ap^r, Cm^r, Sm^r, Sn^r, Km^r
RK2	56.4	Yes	3–5	Ap^r, Km^r, Tc^r

Note: Antibiotic abbreviations are as follows: Ap, ampicillin; Cm, chloramphenicol; Km, kanamycin; Sm, streptomycin; Sn, sulphonamide; Tc, tetracycline. $E1^{imm}$ and $DF13^{imm}$ represent immunity to the homologous but not to the heterologous colicin. Thus plasmid ColE1 is resistant to the effects of its own colicin (E1), but not to colicin DF13. Copy number is the number of plasmids per chromosome equivalent.

Source: After Winnacker (1987), *From Genes to Clones*, VCH. Data from Helsinki (1979), *Critical Reviews in Biochemistry* **7**, 83–101, copyright (1979) CRC Press, Inc., Boca Raton, Florida; Kahn *et al.* (1979), *Methods in Enzymology* **68**, 268–280, copyright (1979) Academic Press and Thomas (1981), *Plasmid* **5**, 10–19, copyright (1981) Academic Press. Reproduced with permission.

features. These will not be described here; details can be found in the texts listed in Suggestions for further reading. However, one additional pBR322-based plasmid is worthy of inclusion, as it is widely used and has some advantages over its progenitor. The plasmid is pAT153, which is a **deletion derivative** of pBR322 (See Fig. 5.1). The plasmid was isolated by removal of two fragments of DNA from pBR322, using the restriction enzyme *Hae*II. The amount of DNA removed was small (705 base-pairs), but the effect was to increase the copy number some threefold, and to remove sequences necessary for mobilisation. Thus pAT153 is in some respects a 'better' vector than pBR322, being present as more copies per cell and having a greater degree of biological containment because it is not mobilisable.

The presence of two antibiotic resistance genes (Ap^r and Tc^r) enables selection for cells harbouring the plasmid, as such cells will be resistant to both ampicillin and tetracycline. An added advantage is that the unique restriction sites within the antibiotic resistance genes permit selection of

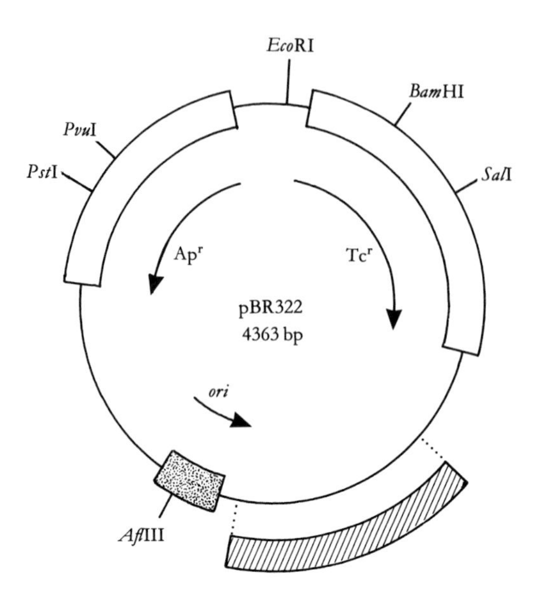

Fig. 5.1. Map of plasmid pBR322. Important regions indicated are the genes for ampicillin and tetracycline resistance (Ap^r and Tc^r) and the origin of replication (*ori*). Some unique restriction sites are given. The hatched region represents the two fragments that were removed from pBR322 to generate pAT153.

cloned DNA by what is known as **insertional inactivation**, where the inserted DNA interrupts the coding sequence of the resistance gene and so alters the phenotype of the cell carrying the recombinant. This is discussed further in section 7.1.2.

5.2.3 More exotic plasmid vectors

Although plasmids pBR322 and pAT153 are widely used for many applications in gene cloning, there are other plasmid vectors available. Generally these have been constructed so that they have particular characteristics not found in the simpler vectors. They may contain specific promoters for the expression of inserted genes, or they may offer other advantages such as direct selection for recombinants. Despite these advantages, the well-tried vectors such as pBR322 and pAT153 are often more than sufficient for the experimental procedure that is being performed.

One series of plasmid vectors that has proved popular is the pUC family. These plasmids have a region that contains several unique restriction

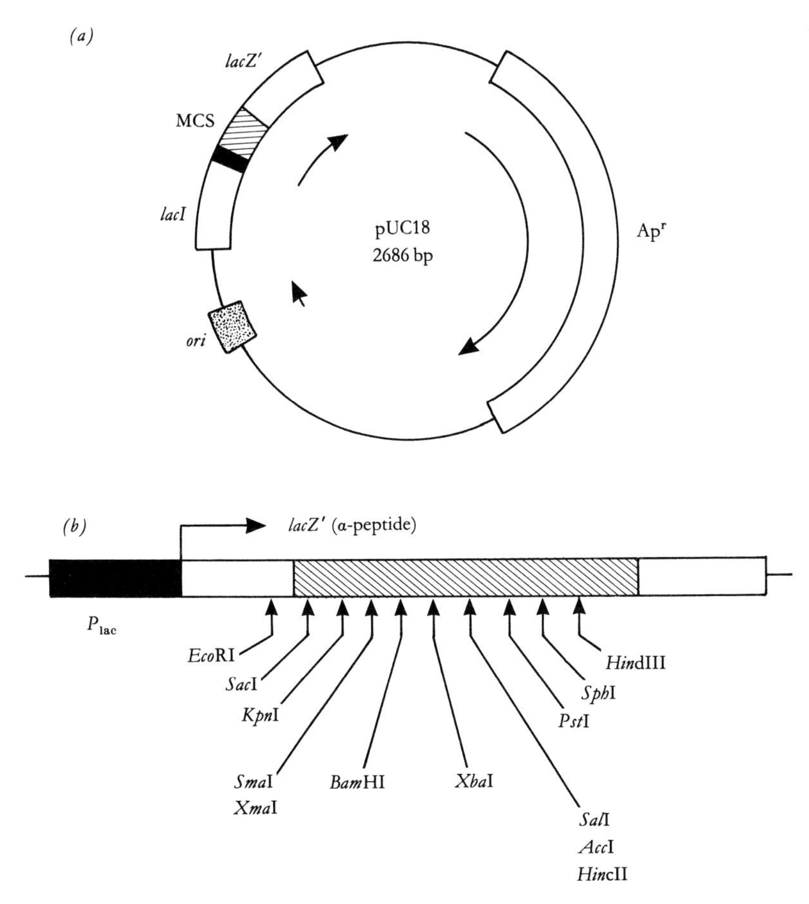

Fig. 5.2. Map of plasmid pUC18. (*a*) The physical map, with the positions of the origin of replication (*ori*) and the ampicillin resistance gene (Ap^r) indicated. The *lacI* gene (lac repressor), multiple cloning site (MCS) or polylinker and the *lacZ'* gene (α-peptide fragment of ß-galactosidase) are also shown. (*b*) The polylinker region. This has multiple restriction sites immediately downstream from the lac promoter (P_{lac}). The in-frame insert used to create the MCS is hatched. Plasmid pUC19 is identical with pUC18 apart from the orientation of the polylinker region, which is reversed.

endonuclease sites in a short stretch of the DNA. This region is known as a **polylinker** or **multiple cloning site** (MCS), and is useful because of the choice of site available for insertion of DNA fragments during recombinant production. A map of one of the pUC vectors, with the restriction sites in its polylinker region, is shown in Fig. 5.2. In addition to the multiple cloning sites in the polylinker region, the pUC plasmids have a region of the ß-galactosidase gene that codes for what is known as the α-peptide. This sequence contains the polylinker region, and insertion of a DNA fragment into one of the cloning sites results in a non-functional α-peptide. This forms the basis for a powerful direct recombinant screening method using the chromogenic substrate X-gal, as outlined in sections 7.1.1 and 7.1.2.

Although plasmid vectors have many useful properties and are essential for gene manipulation, they do have a number of disadvantages. One of the major drawbacks is the size of DNA fragment that can be inserted into plasmids, the maximum being around 5 kb of DNA before cloning efficiency is severely affected. In many cases this is not a problem, but in some applications it is important to maximise the size of fragments that may be cloned. Such a case is the generation of a **genomic library**, in which all the sequences present in the genome of an organism are represented. For this type of approach, vectors that can accept larger pieces of DNA are required. Examples of suitable vectors are those based on bacteriophage lambda (λ); these are considered in the next section.

5.3 Bacteriophage vectors for use in *E. coli*

Although bacteriophage-based vectors are in many ways more specialised than plasmid vectors, they fulfil essentially the same function, i.e. they act as carrier molecules for fragments of DNA. Two types of bacteriophage (λ and M13) have been extensively developed for cloning purposes; these will be described to illustrate the features of bacteriophages and the vectors derived from them.

5.3.1 What are bacteriophages?

In the 1940s Max Delbrück, and the 'Phage Group' that he brought into existence, laid the foundations of modern molecular biology by studying bacteriophages. These are literally 'eaters of bacteria', and are viruses that are dependent on bacteria for their propagation. The term bacteriophage is often shortened to **phage**, and can be used to describe either one or many particles of the same type. Thus we might say that a test tube contained one λ phage or 2×10^6 λ phage particles. The plural term **phages** is used when different types of phage are being considered; we therefore talk of T4, M13 and λ as being phages.

Structurally, phages fall into three main groups: (i) tailless, (ii) head with tail and (iii) filamentous. The genetic material may be single or double-

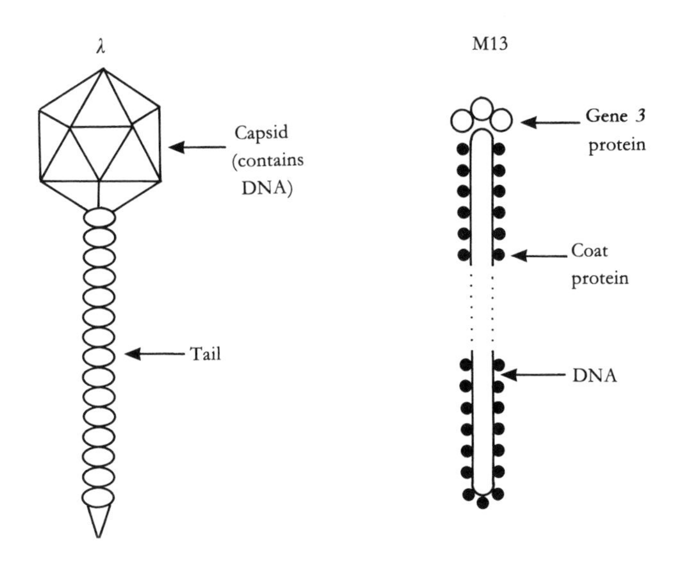

Fig. 5.3. Structure of bacteriophages λ and M13. Phage λ has a capsid or head which encloses the double-stranded DNA genome. The tail region is required for adsorption to the host cell. M13 has a simpler structure, with the single-stranded DNA genome being enclosed in a protein coat. The gene *3* product is important in both adsorption and extrusion of the phage. M13 is not drawn to scale; in reality it is a long thin structure.

stranded DNA or RNA, with double-stranded DNA (dsDNA) being found most often. In tailless and tailed phages the genome is encapsulated in an icosahedral protein shell called a **capsid** (sometimes known as a phage coat or head). In typical dsDNA phages, the genome makes up about 50% of the mass of the phage particle. Thus phages represent relatively simple systems when compared to bacteria, and for this reason they have been extensively used as models for the study of gene expression. The structure of phages λ and M13 is shown in Fig. 5.3.

Phages may be classified as either **virulent** or **temperate**, depending on their life cycles. When a phage enters a bacterial cell it can produce more phage and kill the cell (this is called the **lytic** growth cycle), or it can integrate into the chromosome and remain in a quiescent state without killing the cell (this is the **lysogenic** cycle). Virulent phages are those that exhibit a lytic life cycle only. Temperate phages exhibit lysogenic life cycles, but most can also undergo the lytic response when conditions are suitable. The best-known example of a temperate phage is λ, which has been the subject of intense research effort and is now more or less fully characterised in terms of its structure and mode of action.

The genome of phage λ is 48.5 kb in length, and encodes some 46 genes

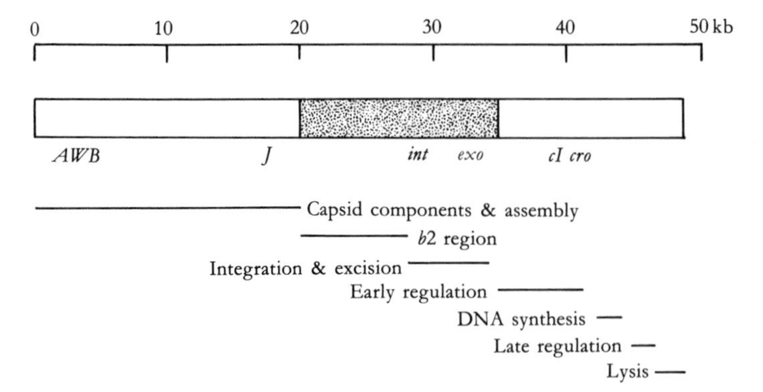

Fig. 5.4. Map of the phage λ genome. Some of the genes are indicated. Functional regions are shown by horizontal lines and annotated. The non-essential region that may be manipulated in vector construction is shaded.

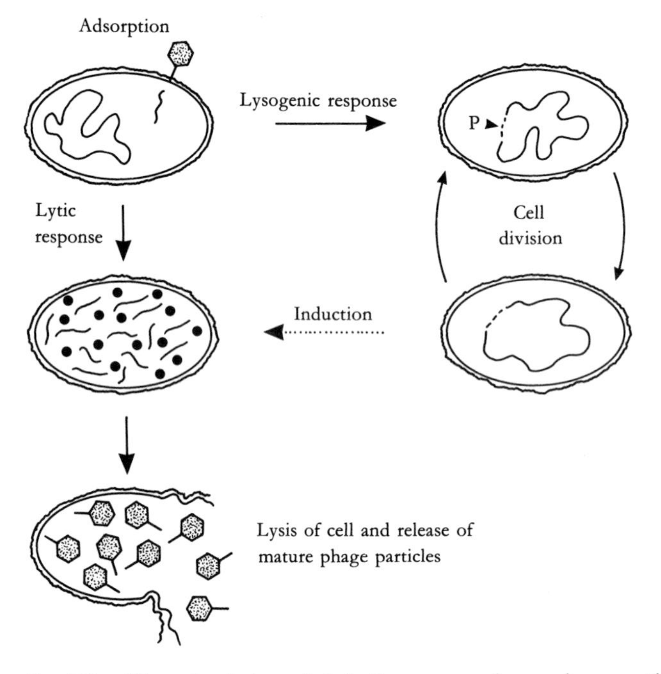

Fig. 5.5. Life cycle of phage λ. Infection occurs when a phage particle is adsorbed and the DNA injected into the host cell. In the lytic response, the phage takes over the host cell and produces copies of the phage genome and structural proteins. Mature phage particles are then assembled and released by lysis of the host cell. In the lysogenic response the phage DNA integrates into the host genome as a prophage (P), which can be maintained through successive cell divisions. The lytic response can be induced in a lysogenic bacterium in response to a stimulus such as ultraviolet light.

(Fig. 5.4). The entire genome has been sequenced (this in itself represents one of the milestones of molecular genetics), and all the regulatory sites are known. At the ends of the linear genome there are short (12 bp) single-stranded regions that are complementary. These act as cohesive or 'sticky' ends, which enable circularisation of the genome following infection. The region of the genome that is generated by the association of the cohesive ends is known as the *cos* site.

Phage infection begins with **adsorption**, which involves the phage particle binding to receptors on the bacterial surface (Fig. 5.5). When the phage has adsorbed, the DNA is injected into the cell and the life cycle can begin. The genome circularises and the phage initiates either the lytic or lysogenic cycle, depending on a number of factors that include the nutritional and metabolic state of the host cell and the **multiplicity of infection** (m.o.i. – the ratio of phage to bacteria during adsorption). If the lysogenic cycle is initiated, the phage genome integrates into the host chromosome and is maintained as a **prophage**. It is then replicated with the chromosomal DNA and passed on to daughter cells in a stable form. If the lytic cycle is initiated, a complex sequence of transcriptional events essentially enables the phage to take over the host cell and produce multiple copies of the genome and the structural proteins. These components are then assembled or **packaged** into mature phage, which are released following lysis of the host cell.

To determine the number of bacteriophage present in a suspension, serial dilutions of the phage stock are mixed with an excess of indicator bacteria (m.o.i. is very low) and plated onto agar using a soft agar overlay. On incubation, the bacteria will grow to form what is termed a bacterial **lawn**. Phage that grow in this lawn will cause lysis of the cells that the phage infects, and as this growth spreads a cleared area or **plaque** will develop (Fig. 5.6). Plaques can then be counted to determine the number of **plaque forming units** (p.f.u.) in the stock suspension, and may be picked from the plate for further growth and analysis. Phage may be propagated in liquid culture by infecting a growing culture of the host cell and incubating until cell lysis is complete, the yield of phage particles depending on the multiplicity of infection and the stage in the bacterial growth cycle at which infection occurs.

The filamentous phage M13 differs from λ both structurally (Fig. 5.3) and in its life cycle. The M13 genome is a single-stranded circular DNA molecule 6407 bp in length. The phage will infect only *E. coli* that have F-pili (threadlike protein 'appendages' found on conjugation-proficient cells), although the mechanism by which the M13 genome enters the

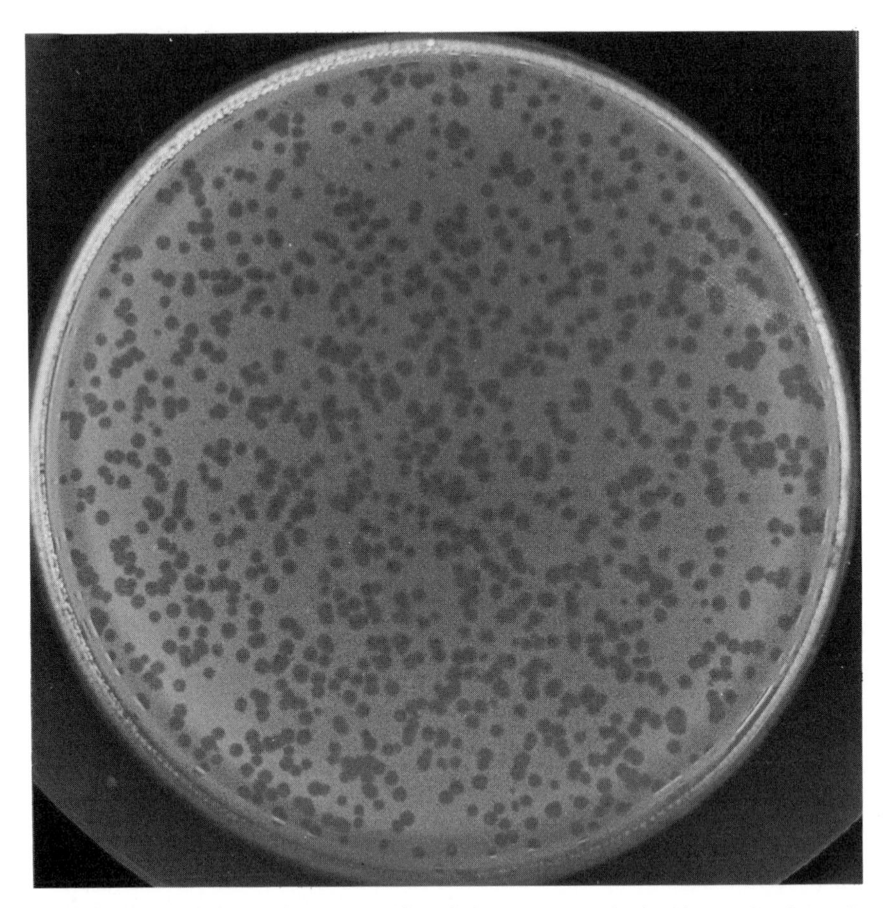

Fig. 5.6. Bacteriophage plaques. Particles of phage λ were mixed with a strain of *E. coli* and plated using a soft agar overlay. After overnight incubation the bacterial cells grow to form a lawn, in which regions of phage infection appear as cleared areas or plaques. Photograph courtesy of Dr M. Stronach.

bacterial cell is still not fully understood. When the DNA enters the cell, it is converted to a double-stranded molecule known as the **replicative form** or **RF**, which replicates until there are about 100 copies in the cell. At this point DNA replication becomes asymmetric, and single-stranded copies of the genome are produced and extruded from the cell as M13 particles. The bacterium is not lysed and remains viable during this process, although growth and division are slower than in non-infected cells.

5.3.2 Vectors based on bacteriophage λ

The utility of phage λ as a cloning vector depends on the fact that not all of the λ genome is essential for the phage to function. Thus there is scope for the introduction of exogenous DNA, although certain requirements have had to be met during the development of cloning vectors based on phage λ. Firstly, the arrangement of genes on the λ genome will determine which parts can be removed or replaced for the addition of exogenous DNA. It is fortunate that the central region of the λ genome (between positions 20 and 35 on the map shown in Fig. 5.4) is largely dispensable, so no complex rearrangement of the genome *in vitro* is required. The central region controls mainly the lysogenic properties of the phage, and much of this region can be deleted without impairing the functions required for the lytic infection cycle. Secondly, wild-type λ phage will generally have multiple recognition sites for the restriction enzymes commonly used in cloning procedures. This can be a major problem, as it limits the choice of sites for the insertion of DNA. In practice, it is relatively easy to select for phage that have reduced numbers of sites for particular restriction enzymes, and the technique of mutagenesis *in vitro* may be used to modify remaining sites that are not required. Thus it is possible to obtain phage that have the desired combination of restriction enzyme recognition sites.

One of the major drawbacks of λ vectors is that the capsid places a physical constraint on the amount of DNA that can be incorporated during phage assembly, which limits the size of exogenous DNA fragments that can be cloned. During packaging, viable phage particles can be produced from DNA that is between approximately 38 kb and 51 kb in length. Thus a wild-type phage genome could accommodate only around 2.5 kb of cloned DNA before becoming too large for viable phage production. This limitation has been minimised by careful construction of vectors to accept pieces of DNA that are close to the theoretical maximum for the particular construct. Such vectors fall into two main classes: (i) **insertion** vectors and (ii) **replacement** or **substitution** vectors. The difference between these two types of vector is outlined in Fig. 5.7.

There is a bewildering variety of λ vectors available for use in cloning experiments, each with slightly different characteristics. The choice of vector has to be made carefully, with aspects such as the size of DNA fragments to be cloned and the preferred selection/screening method being taken into account. To illustrate the structural characteristics of λ vectors, two insertion and two replacement vectors are described briefly. Functional

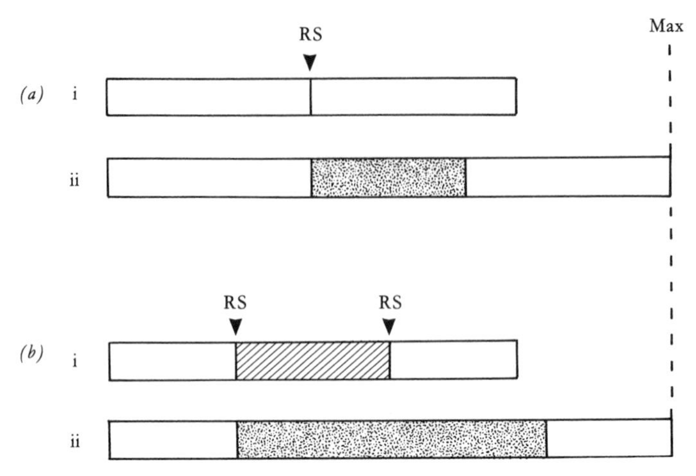

Fig. 5.7. Insertion and replacement phage vectors. (*a*) An insertion vector is shown in part i. Such vectors have a single restriction site (RS). To generate a recombinant DNA is inserted into this site. The size of fragment that may be cloned is therefore determined by the difference between the vector size and the maximum packagable fragment size (Max). Insert DNA is shaded in part ii. (*b*) A replacement vector is shown in part i. These vectors have two restriction sites (RS) which flank a region known as the stuffer fragment (hatched). Thus a section of the phage genome is replaced during cloning into this site, as shown in part ii. This approach enables larger fragments to be cloned than is possible with insertion vectors.

aspects are discussed in Chapter 7 when selection and screening methods are considered.

Insertion vectors have a single recognition site for one or more restriction enzymes, which enables DNA fragments to be inserted into the λ genome. Examples of insertion vectors include λgt10 and Charon 16A. The latter is one of a series of vectors named after the ferryman of Greek mythology, who conveyed the spirits of the dead across the River Styx – a rather apt example of what we might call 'bacteriophage culture'! These two insertion vectors are illustrated in Fig. 5.8. Each has a single *Eco*RI site into which DNA can be inserted. In λgt10 (43.3 kb) this generates left and right 'arms' of 32.7 and 10.6 kb respectively, which can in theory accept insert DNA fragments up to approximately 7.6 kb in length. The *Eco*RI site lies within the *cI* gene (λ repressor), and this forms the basis of a selection/screening method based on plaque formation and morphology (see section 7.1.2). In Charon 16A (41.8 kb), the arms generated by *Eco*RI digestion are 19.9 kb (left arm) and 21.9 kb (right arm), and fragments of up to approximately 9 kb may be cloned. The *Eco*RI site in Charon 16A lies

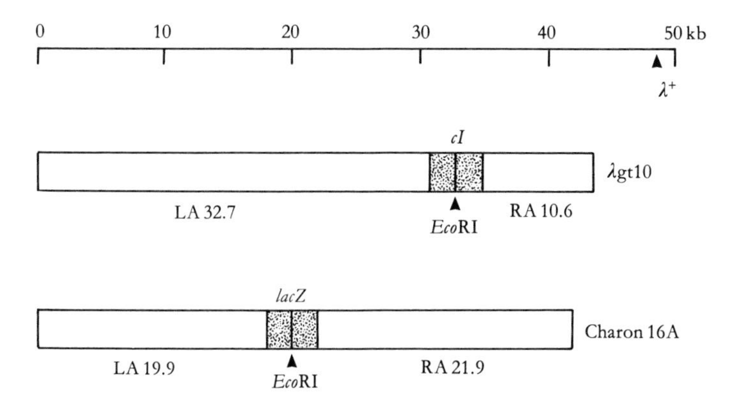

Fig. 5.8. Bacteriophage λ insertion vectors λgt10 and Charon 16A. The *cI* and *lacZ* genes in λgt10 and Charon 16A, respectively, are shaded. Within these genes there is an *Eco*RI site for cloning into. The lengths of the left and right arms (LA and RA, in kb) are given. The size of the wild-type λ genome is marked on the scale bar as λ^+.

within the ß-galactosidase gene (*lacZ*), which enables the detection of recombinants using X-gal (see section 7.1.2).

Insertion vectors offer limited scope for cloning large pieces of DNA, and thus replacement vectors were developed in which a central 'stuffer' fragment is removed and replaced with the insert DNA. Two examples of λ replacement vectors are EMBL4 and Charon 40 (Fig. 5.9). EMBL4 (41.9 kb) has a central 13.2 kb stuffer fragment flanked by inverted polylinker sequences containing sites for the restriction enzymes *Eco*RI, *Bam*HI and *Sal*I. Two *Sal*I sites are also present in the stuffer fragment. DNA may be inserted into any of the cloning sites, the choice depending on the method of preparation of the fragments. Often a partial *Sau*3A or *Mbo*I digest is used in the preparation of a genomic library (see section 6.3.2), which enables insertion into the *Bam*HI site. Such inserts may be released from the recombinant by digestion with *Eco*RI. During preparation of the vector for cloning, the *Bam*HI digestion (which generates sticky ends for accepting the insert DNA) is often followed by a *Sal*I digestion. This cleaves the stuffer fragment at the two internal *Sal*I sites and also releases short *Bam*HI/*Sal*I fragments from the polylinker region. This is helpful because it prevents the stuffer fragment from re-annealing with the left and right arms and generating a viable phage that is non-recombinant.

DNA fragments between approximately 9 and 22 kb may be cloned in EMBL4, the lower limit representing the minimum size required to form viable phage particles (left arm + insert + right arm must be greater than 38 kb) and the upper the maximum packagable size of around 51 kb. These

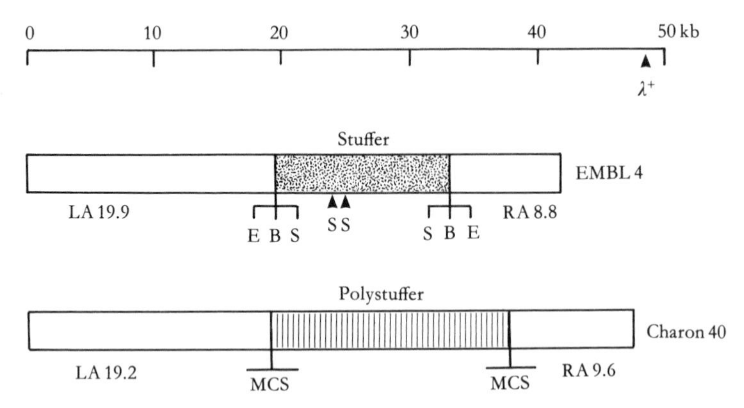

Fig. 5.9. Bacteriophage λ replacement vectors EMBL4 and Charon 40. The stuffer fragment in EMBL4 is 13.2 kb, and is flanked by inverted polylinkers containing the sites for *Eco*RI (E), *Bam*HI (B) and *Sal*I (S). In Charon 40 the polystuffer is composed of short repeated regions that are cleaved by *Nae*I. The multiple cloning site (MCS) in Charon 40 carries a wider range of restriction sites than that in EMBL4.

size constraints can act as a useful initial selection method for recombinants, although an additional genetic selection mechanism can be employed with EMBL4 (the Spi^- phenotype, see section 7.1.4).

Charon 40 is a replacement vector in which the stuffer fragment is composed of multiple repeats of a short piece of DNA. This is known as a **polystuffer**, and it has the advantage that the restriction enzyme *Nae*I will cut the polystuffer into its component parts. This enables efficient removal of the polystuffer during vector preparation, and most of the surviving phage will be recombinant. The polystuffer is flanked by polylinkers with a more extensive range of restriction sites than found in EMBL4, which increases the choice of restriction enzymes that may be used to prepare the insert DNA. The size range of fragments that may be cloned in Charon 40 is similar to that for EMBL4.

5.3.3 Vectors based on bacteriophage M13

Two aspects of M13 infection are of value to the genetic engineer. Firstly, the RF is essentially similar to a plasmid, and can be isolated and manipulated using the same techniques. A second advantage is that the single-stranded DNA produced during the infection is useful in techniques such as DNA sequencing by the dideoxy method (see section 3.6.2). This aspect alone made M13 immediately attractive as a potential vector.

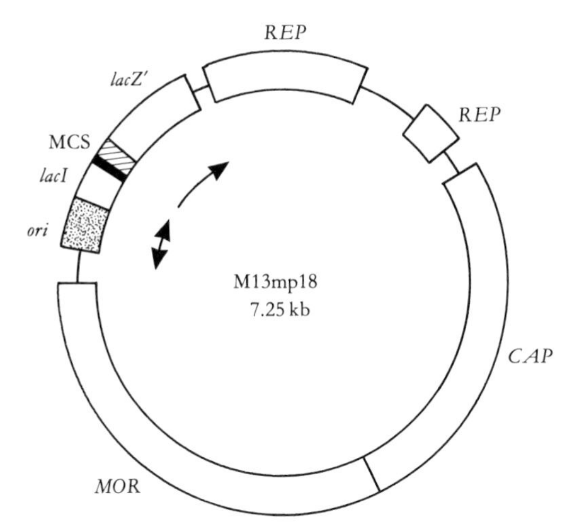

Fig. 5.10. Map of the filamentous phage vector M13mp18. The double-stranded replicative form is shown. The polylinker region (MCS) is the same as that found in plasmid pUC18 (Fig. 5.2). Genes in the *REP* region encode proteins that are important for DNA replication. The *CAP* and *MOR* regions contain genes that specify functions associated with capsid formation and phage morphogenesis, respectively. The vector M13mp19 is identical except for the orientation of the polylinker region.

Unlike phage λ, M13 does not have any non-essential genes. The 6407 bp genome is also used very efficiently in that most of it is taken up by gene sequences, so that the only part available for manipulation is a 507 bp intergenic region. This has been used to construct the M13mp series of vectors, by inserting a polylinker/*lacZ* α-peptide sequence into this region (Fig. 5.10). This enables the X-gal screening system to be used for the detection of recombinants, as is the case with the pUC plasmids. When M13 is grown on a bacterial lawn, 'plaques' appear due to the reduction in growth of the host cells (which are not lysed), and these may be picked for further analysis.

A second disadvantage of M13 vectors is the fact that they do not function efficiently when long DNA fragments are inserted into the vector. Although in theory there should be no limit to the size of clonable fragments, as the capsid structure is determined by the genome size (unlike phage λ), there is a marked reduction in cloning efficiency with fragments longer than about 1.5 kb. In practice this is not a major problem, as the main use of M13 vectors is in sub-cloning small DNA fragments for sequencing. In this application single-stranded DNA production, coupled with ease of purification of the DNA from the cell culture, outweighs any size limitation, although this has also been alleviated by the construction of hybrid plasmid/M13 vectors (see section 5.4.1).

5.4 Other vectors

So far I have concentrated on plasmid and bacteriophage vectors for use in *E. coli* hosts, which still represents a major part of the technology of gene manipulation. However, there are other host/vector systems that are just as important, although they may be more specialised. In this section I examine the features of some additional bacterial vectors, and some vectors for use in other organisms.

5.4.1 Hybrid plasmid/phage vectors

One feature of phage vectors is that the technique of packaging *in vitro* (see section 5.5.2 for details) is sequence-independent, apart from the requirement of having the *cos* sites separated by DNA of packagable size (38–51 kb). This has been exploited in the construction of vectors that are made up of plasmid sequences joined to the *cos* sites of phage λ. Such vectors are known as **cosmids**. They are small (4–6 kb) and can therefore accommodate cloned DNA fragments up to some 47 kb in length. As they lack phage genes, they behave as plasmids when introduced into *E. coli* by the packaging/infection mechanism of λ. Cosmid vectors therefore offer an apparently ideal system – a highly efficient and specific method of introducing the recombinant DNA into the host cell, and a cloning capacity some twofold greater than the best λ replacement vectors. However, they are not without disadvantages, and often the gains of using cosmids instead of phage vectors are offset by losses in terms of ease of use and further processing of cloned sequences.

Hybrid plasmid/phage vectors in which the phage functions are expressed and utilised in some way are known as **phasmids**. One such vector is λZAP, which is a complex λ insertion vector that enables cloned DNA fragments to be excised *in vivo* as part of a plasmid. This automatic excision is useful in that it removes the need to sub-clone inserts from λ into plasmid vectors for further manipulations.

Hybrid plasmid/phage vectors have been developed to overcome the size limitation of the M13 cloning system, and are now widely used for applications such as DNA sequencing and the production of probes for use in hybridisation studies. These vectors are essentially plasmids, which contain the f1 (M13) phage origin of replication. When cells containing the plasmid are superinfected with phage, they produce single-stranded copies of the plasmid DNA and secrete these into the medium as M13-like particles. Vectors such as pEMBL9 or pBluescript can accept DNA fragments of up to 10 kb.

5.4.2 Vectors for use in eukaryotic cells

When eukaryotic host cells are considered, vector requirements become more complex. Bacteria are relatively simple in genetic terms, whereas eukaryotic cells have multiple chromosomes that are held within the membrane-bound nucleus. Given the wide variety of eukaryotes, it is not surprising that vectors tend to be highly specialised and designed for specific purposes.

A number of vectors for use in yeast cells have been developed, with the choice of vector depending on the particular application. **Yeast episomal plasmids** (YEps) are based on the naturally occurring yeast 2 μm plasmid, and may replicate autonomously or integrate into a chromosomal location. **Yeast integrative plasmids** (YIps) are designed to integrate into the chromosome in a similar way to the YEp plasmids, and **yeast replicative plasmids** (YRps) remain as independent plasmids and do not integrate. Plasmids that contain sequences from around the centromeric region of chromosomes are known as **yeast centromere plasmids** (YCps), and these behave essentially as mini-chromosomes. Finally, the **yeast artificial chromosomes** (YACs) represent the most sophisticated yeast vectors (Fig. 5.11). These have centromeric and telomeric regions, and can be used for cloning very large pieces of DNA, because the recombinant is maintained essentially as a yeast chromosome. In fact, stability of YACs appears to increase with increasing size of the cloned fragment, unlike many other vector systems.

When dealing with higher eukaryotes that are multicellular, such as plants and animals, the problems of introducing recombinant DNA into the organism become slightly different than those which apply to microbial eukaryotes such as yeast. The aims of genetic engineering in higher eukaryotes are twofold: (i) to introduce recombinant DNA into plant and animal cells in tissue culture, for basic research on gene expression or for the production of useful proteins, and (ii) to alter the genetic makeup of the organism and produce a **transgenic**, in which all the cells will carry the genetic modification. The latter aim in particular can pose technical difficulties, as the recombinant DNA has to be introduced very early in development or in some sort of vector that will promote the spread of the recombinant sequence throughout the organism.

Vectors used for plant and animal cells may be introduced into cells

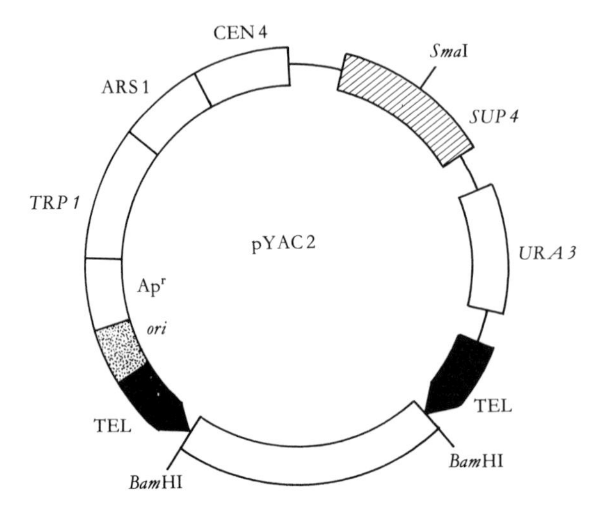

Fig. 5.11. Map of the yeast artificial chromosome vector pYAC2. This carries the origin of replication (*ori*; shaded) and ampicillin resistance gene (Ap^r) from pBR322, and yeast sequences for replication (ARS1) and chromosome structure (centromere, CEN4; and telomeres, TEL). The TEL sequences are separated by a fragment flanked by two *Bam*HI sites. The genes *TRP1* and *URA3* may be used as selectable markers in yeast. The cloning site *Sma*I lies within the *SUP4* gene (hatched). From Kingsman and Kingsman (1988), *Genetic Engineering*, Blackwell. Reproduced with permission.

directly by techniques such as those described in section 5.5.3, or they may have a biological entry mechanism if based on viruses or other infectious agents such as agrobacteria. Some examples of vectors for plant and animal cells are given in Table 5.3. Their use is described further when considering the production of transgenics in Chapter 8.

5.5 Getting DNA into cells

Manipulation of vector and insert DNAs to produce recombinant molecules is carried out in the test-tube, and we are then faced with the task of getting the recombinant DNA into the host cell for propagation. The efficiency of this step is often a crucial factor in determining the success of a given cloning experiment, particularly when a large number of recombinants is required. The methods available depend on the type of host/vector system, and range from very simple procedures to much more complicated and esoteric ones. In this section I look at some of the methods available for getting recombinant DNA into host cells.

Table 5.3. *Some possible vectors for plant and animal cells*

Cell type	Vector type	Genome	Examples
Plant cells	Plasmid	DNA	Ti plasmids of *Agrobacterium tumefaciens*
	Viral	DNA	Cauliflower mosaic virus, Geminiviruses
		RNA	Tobacco mosaic virus
Animal cells	Plasmid	DNA	Various types of plasmid vector are available. Many are hybrid vectors containing part of the SV40 genome
	Viral	DNA	Baculoviruses Papilloma viruses Simian virus 40 (SV40) Vaccinia virus
	Viral	RNA	Retroviruses
	Transposon	DNA	P elements in *Drosophila melanogaster*

Note: In many cases the 'vectorology' associated with a particular group of potential vectors is not well advanced, and often well-tried vectors continue to be used and developed further for particular applications.

5.5.1 Transformation and transfection

The techniques of transformation and transfection represent the simplest methods available for getting recombinant DNA into cells. In the context of cloning in *E. coli* cells, transformation refers to the uptake of plasmid DNA, and transfection to the uptake of phage DNA. Transformation is also used more generally to describe uptake of any DNA by any cell, and can also be used in a different context when talking about a **growth transformation** such as occurs in the production of a cancerous cell.

Transformation in bacteria was first demonstrated in 1928 by Frederick Griffith, in his famous 'transforming principle' experiment that paved the way for the discoveries that eventually showed that genes were made of DNA. However, not all bacteria can be transformed easily, and it was not until the early 1970s that transformation was demonstrated in *E. coli*, the mainstay of gene manipulation technology. To effect transformation of *E. coli*, the cells need to be made **competent**. This is achieved by soaking the cells in an ice-cold solution of calcium chloride, which induces competence in a way that is still not fully understood. Transformation of competent cells is carried out by mixing the plasmid DNA with the cells, incubating on ice for 20–30 min, and then giving a brief heat shock (2 min at 42 °C is often used) which appears to enable the DNA to enter the cells. The transformed cells are usually incubated in a nutrient broth at 37 °C for 60–90 min to enable the plasmids to become established and permit phenotypic expression of their traits. The cells can then be plated out onto selective media for propagation of cells harbouring the plasmid.

Transformation is an inefficient process in that only a very small percentage of competent cells become transformed, representing uptake of a fraction of the plasmid DNA that is available. Thus the process can become the critical step in a cloning experiment where a large number of individual recombinants is required, or when the starting material is limiting. Despite these potential disadvantages, transformation is an essential technique, and with care can yield up to 10^9 transformed cells (**transformants**) per microgram of input DNA, although transformation frequencies of around 10^6 or 10^7 transformants per microgram are more often achieved in practice. Transfection is a similar process to transformation, the difference being that phage DNA is used instead of plasmid DNA. It is again a somewhat inefficient process, and it has largely been superseded by packaging *in vitro* for applications that require the introduction of phage DNA into *E. coli* cells.

5.5.2 Packaging phage DNA *in vitro*

During the lytic cycle of phage λ, the phage DNA is replicated to form what is known as a **concatemer**. This is a very long DNA molecule composed of many copies of the λ genome, linked together by the *cos* sites (Fig. 5.12(*a*)). When the phage particles are assembled the DNA is packaged into the capsid, which involves cutting the DNA at the *cos* sites using a phage-encoded endonuclease. Mature phage particles are thus produced, ready to be released on lysis of the cell, and capable of infecting other cells. This process normally occurs *in vivo*, the particular functions being encoded by the phage genes. However, it is possible to carry out the process in the test tube, which enables recombinant DNA that is generated as a concatemer to be packaged into phage particles.

To enable packaging *in vitro*, the components of the λ capsid, and the endonuclease, must be available. In practice, two strains of bacteria are used to produce a lysate known as a **packaging extract**. Each strain is mutant in one function of phage morphogenesis, so that the packaging extracts will not work in isolation. When the two are mixed with the concatemeric recombinant DNA under suitable conditions, all the components are

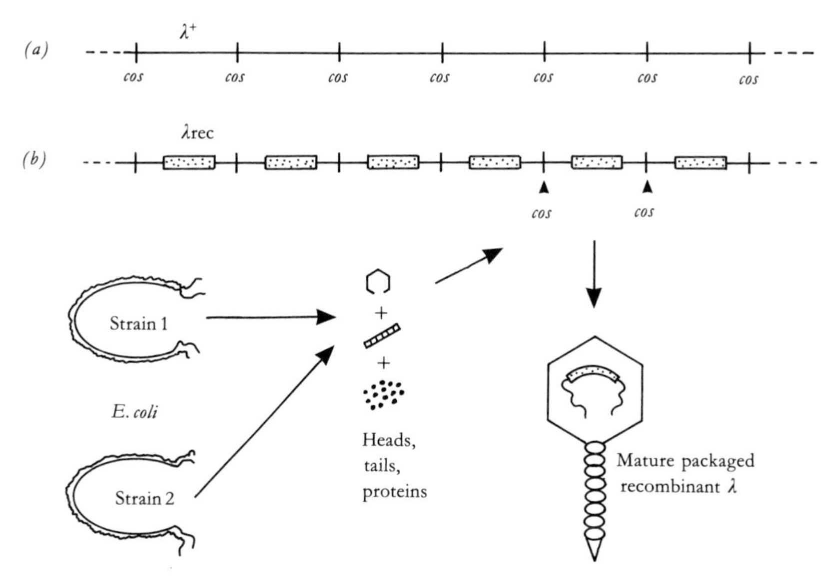

Fig. 5.12. Phage DNA and packaging. (*a*) A concatemeric DNA molecule composed of wild-type phage DNA (λ^+). The individual genomes are joined at the *cos* sites. (*b*) Recombinant genomes (λrec) are shown being packaged *in vitro*. A mixed lysate from two bacterial strains supplies the head and tail precursors and the proteins required for the formation of mature λ particles. On adding this mixture to the concatemer, the DNA is cleaved at the *cos* sites (arrowed) and packaged into individual phage particles, each containing a recombinant genome.

available and phage particles are produced. These particles can then be used to infect *E. coli* cells, which are plated out to obtain plaques. The process of packaging *in vitro* is summarised in Fig. 5.12(*b*).

5.5.3 Alternative DNA delivery methods

The methods available for introducing DNA into bacterial cells are not easily transferred to other cell types. The phage-specific packaging system is not available for other systems, and transformation by normal methods may prove impossible or too inefficient to be a realistic option. However, there are alternative methods for introducing DNA into cells. Often these are more technically demanding and less efficient than the bacterial methods, but reliable results have been achieved in many situations where there appeared to be no hope of getting recombinant DNA molecules into the desired cell.

Most of the problems associated with getting DNA into non-bacterial

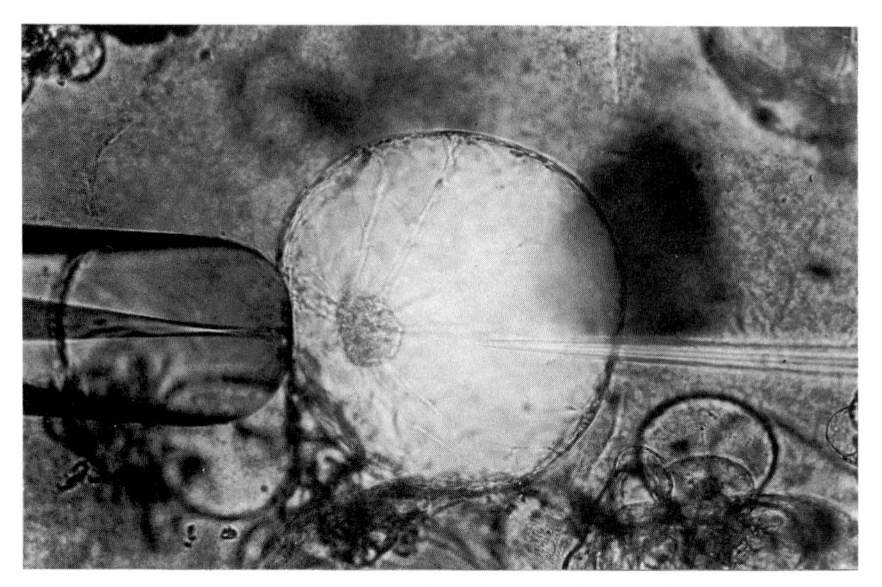

Fig. 5.13. Microinjection of a protoplast-derived potato cell. The cell is held on a glass capillary (on the left of the photograph) by gentle suction. The microinjection needle is made by drawing a heated glass capillary out to a fine point. Using a micromanipulator the needle has been inserted into the cell (on the right of the photograph), where its tip can be seen approaching the cell nucleus. Photograph courtesy of Dr K. Ward.

cells have involved plant cells. Animal cells are relatively flimsy, and can be transformed readily. However, plant cells pose the problem of a rigid cell wall, which is a barrier to DNA uptake. This can be alleviated by the production of **protoplasts**, in which the cell wall is removed enzymatically. The protoplasts can then be transformed using a technique such as **electroporation**, where an electrical pulse is used to create transient holes in the cell membrane, through which the DNA can pass. The protoplasts can then be regenerated. In addition to this application, protoplasts also have an important role to play in the generation of hybrid plant cells by fusing protoplasts together.

An alternative to transformation procedures is to introduce DNA into the cell by some sort of physical method. One way of doing this is to use a very fine needle and inject the DNA directly into the nucleus. This technique is called **microinjection** (Fig. 5.13), and has been used successfully with both plant and animal cells. The cell is held on a glass tube by mild suction and the needle used to pierce the membrane. The technique requires a mechanical micromanipulator and a microscope, and plenty of practice!

A recent and somewhat bizarre development has proved extremely

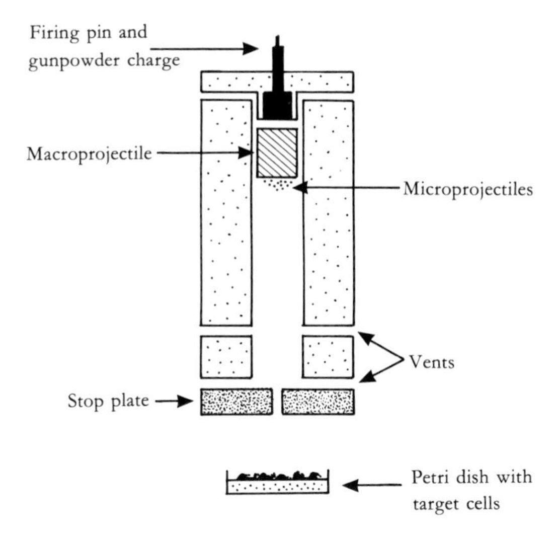

Fig. 5.14. Biolistic apparatus. The DNA is coated onto microprojectiles, which are accelerated by the macroprojectile on firing the gun. At the stop plate the macroprojectile is retained in the chamber and the microprojectiles carry on to the target tissue. Other versions of the apparatus, driven by compressed gas instead of a gunpowder charge, are available.

useful in transformation of plant cells. The technique, which is called **biolistic** DNA delivery, involves literally shooting DNA into cells (Fig. 5.14). The DNA is used to coat microscopic tungsten particles known as **microprojectiles**, which are then accelerated on a **macroprojectile** by firing a gunpowder charge. At one end of the 'gun' there is a small aperture that stops the macroprojectile but allows the microprojectiles to pass through. When directed at cells, these microprojectiles carry the DNA into the cell and, in some cases, stable transformation will occur.

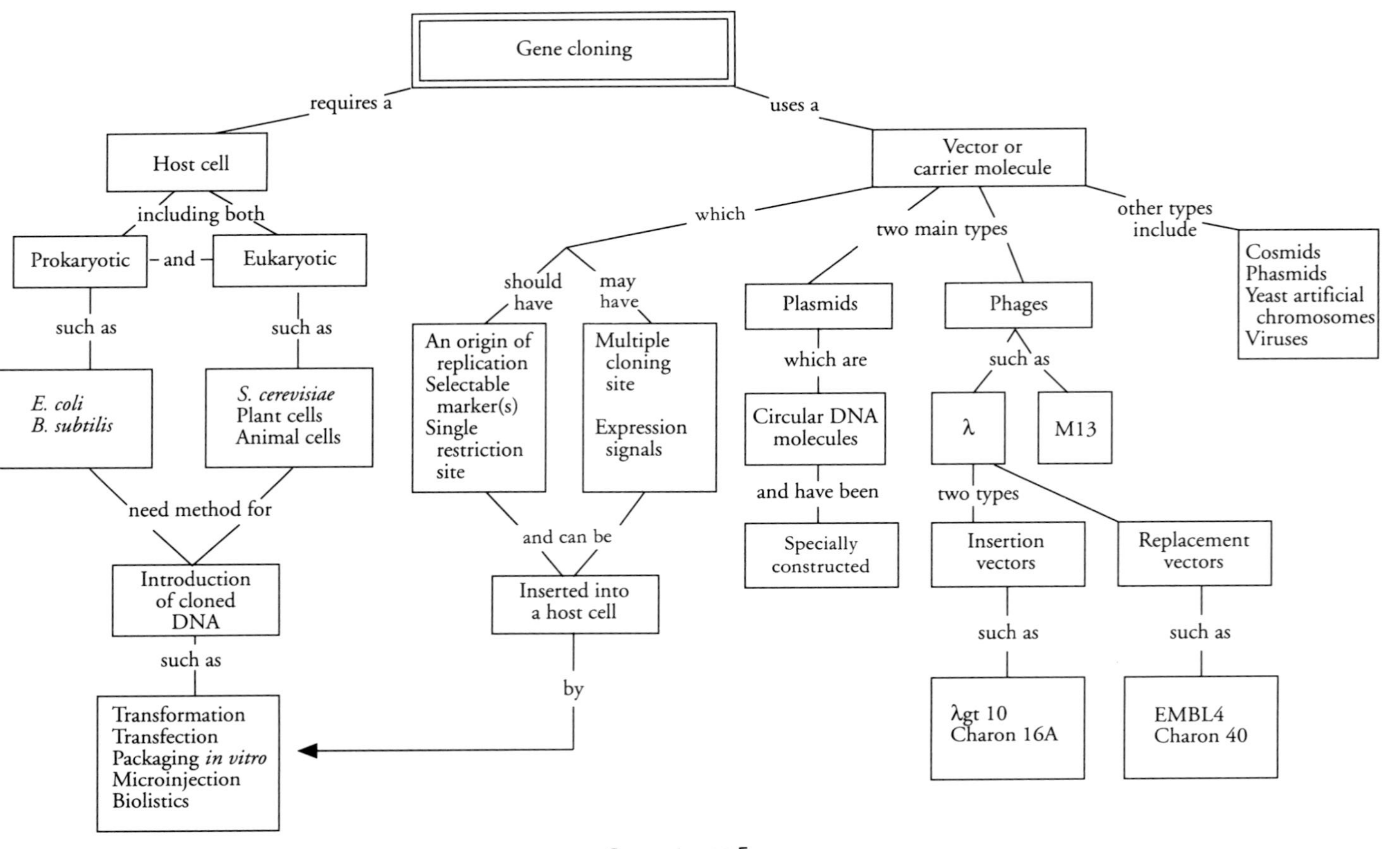

Concept map 5.

6

Cloning strategies

In the previous two chapters I examined the two essential components of genetic engineering, these being (i) the ability to cut, modify and join DNA molecules *in vitro*, and (ii) the host/vector systems that allow recombinant DNA molecules to be propagated. With these components at his or her disposal, the genetic engineer has to devise a cloning strategy that will enable efficient use of the technology to achieve the aims of the experiment. In Chapter 1 I showed that there are basically four stages to any cloning experiment (Fig. 1.1), involving **generation** of DNA fragments, **joining** to a vector, **propagation** in a host cell, and **selection** of the required sequence. In this chapter I examine some of the strategies that are available for completing the first three of these stages, largely restricting the discussion to cloning eukaryotic DNA in *E. coli*. Selection of cloned sequences is discussed in Chapter 7, although the type of selection method that will be used does have to be considered when choosing host/vector combinations for a particular cloning exercise.

6.1 Which approach is best?

The complexity of any cloning experiment depends largely on two factors: (i) the overall aims of the work, and (ii) the type of source material from which the nucleic acids will be isolated for cloning. Thus a strategy to isolate and sequence a relatively small DNA fragment from *E. coli* will be different (and will probably involve fewer stages) than a strategy to produce a recombinant protein in a transgenic eukaryotic organism. There is no

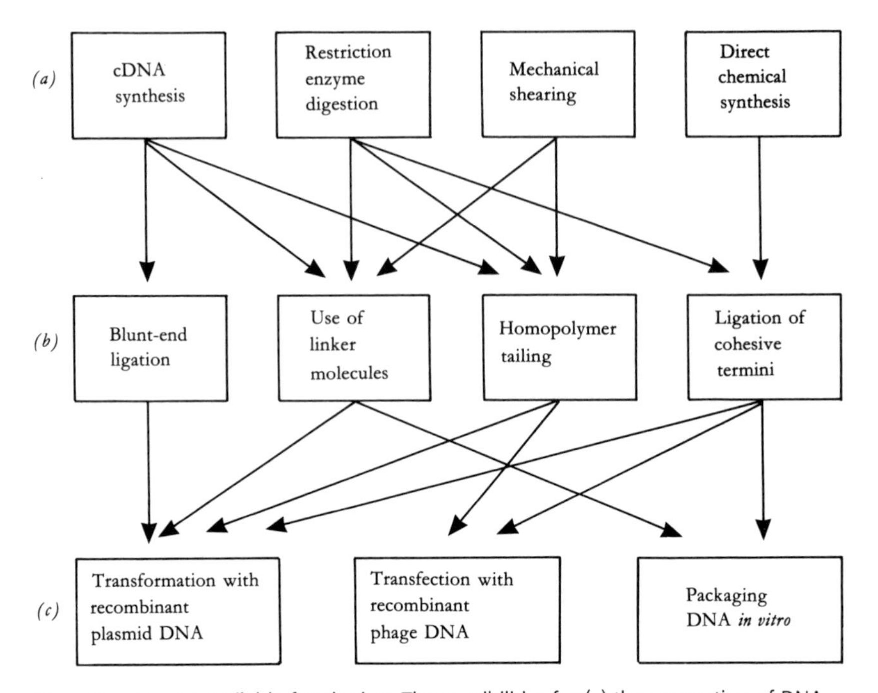

Fig. 6.1. Routes available for cloning. The possibilities for (*a*) the generation of DNA fragments, (*b*) joining to a vector and (*c*) introducing the recombinant DNA into a host cell. Preferred routes are indicated by arrows. Redrawn from Old and Primrose (1989), *Principles of Gene Manipulation*, 4th edition, Blackwell. Reproduced with permission.

single cloning strategy that will cover all requirements. Each project will therefore be unique, and will present its own set of problems that have to be addressed by choosing the appropriate path through the maze of possibilities (see Fig. 6.1). Fortunately, most of the confusion can be eliminated by careful design of experiments and rigorous interpretation of results.

When dealing with eukaryotic organisms, the first major decision is whether to begin with messenger RNA (mRNA) or genomic DNA. Although the DNA represents the complete genome of the organism, it may contain non-coding DNA such as introns, control regions and repetitive sequences. This can sometimes present problems, particularly if the genome is large and the aim is to isolate a single-copy gene. However, if the primary interest is in the control of gene expression, it is obviously necessary to isolate the control sequences, so genomic DNA is the only alternative.

Messenger RNA has two advantages over genomic DNA as a source material. Firstly, it represents the genetic information that is being expressed by the particular cell type from which it is prepared. This can be a very powerful preliminary selection mechanism, as not all the genomic DNA will be represented in the mRNA population. Also, if the gene of interest is highly expressed, this may be reflected in the abundance of its mRNA, and this can make isolation of the clones easier. A second advantage of mRNA is that it, by definition, represents the coding sequence of the gene, with any introns having been removed during RNA processing. Thus production of recombinant protein is much more straightforward if a clone of the mRNA is available.

Although genomic DNA and mRNA are the two main sources of nucleic acid molecules for cloning, it is possible to synthesise DNA *in vitro* if the amino acid sequence of the protein is known. Whilst this is a laborious task for long stretches of DNA, it is a useful technique in some cases, particularly if only short sections of a gene need to be synthesised to complete a sequence prior to cloning.

Having decided on the source material, the next step is to choose the type of host/vector system. Even when cloning in *E. coli* hosts there is still a wide range of strains available, and care must be taken to ensure that the optimum host/vector combination is chosen. When choosing a vector, the method of joining the DNA fragments to the vector and the means of getting the recombinant molecules into the host cell are two main considerations. In practice the host/vector systems in *E. coli* are usually well defined, so it is a relatively straightforward task to choose the best combination, given the type of fragments to be cloned and the desired outcome of the experiment.

In devising a cloning strategy all the points mentioned above have to be considered. Often there will be no ideal solution to a particular problem, and a compromise will have to be accepted. By keeping the overall aim of the experiments in mind, the researcher can minimise the effects of such compromises and choose the most efficient cloning route.

6.2 Cloning from mRNA

Each type of cell in a multicellular organism will produce a range of mRNA molecules. In addition to the expression of general 'housekeeping' genes whose products are required for basic cellular metabolism, cells exhibit tissue-specific gene expression. Thus liver cells, kidney cells, skin cells etc. will each synthesise a different spectrum of tissue-specific proteins (and hence mRNAs). In addition to the **diversity** of mRNAs produced by each

Table 6.1. *mRNA abundance classes*

Source	Number of different mRNAs	Abundance (molecules/cell)
Mouse liver cytoplasmic poly(A)$^+$ RNA	9	12 000
	700	300
	11 500	15
Chick oviduct polysomal poly(A)$^+$ RNA	1	100 000
	7	4000
	12 500	5

Note: The diversity of mRNAs is indicated by the number of different mRNA molecules. There is one mRNA that is present in chick oviduct cells at a very high level (100 000 molecules per cell). This mRNA encodes ovalbumin, the major egg white protein.

Source: After Old & Primrose (1989), *Principles of Gene Manipulation*, 4th edition, Blackwell. Mouse data from Young *et al.* (1976), *Biochemistry* **15**, 2823–2828, copyright (1976) American Chemical Society. Chick data from Axel *et al.* (1976) *Cell* **11**, 247–254, copyright (1976) Cell Press. Reproduced with permission.

cell type, there may well be different **abundance classes** of particular mRNAs. This has important consequences for cloning from mRNA, as it is easier to isolate a specific cloned sequence if it is present as a high proportion of the starting mRNA population. Some examples of mRNA abundance classes are shown in Table 6.1.

6.2.1 Synthesis of cDNA

It is not possible to clone mRNA directly, so it has to be converted into DNA before being inserted into a suitable vector. This is achieved using the enzyme reverse transcriptase (RTase; see section 4.2.2) to produce **complementary DNA** (also known as **copy DNA** or **cDNA**). The classic early method of cDNA synthesis utilises the poly(A) tract at the 3′ end of the mRNA to bind an oligo(dT) primer, which provides the 3′-OH group required by RTase (Fig. 6.2). Given the four dNTPs and suitable conditions, RTase will synthesise a copy of the mRNA to produce a cDNA·mRNA hybrid. The mRNA can be removed by alkaline hydrolysis and the single-stranded (ss) cDNA converted into double-stranded (ds) cDNA by using a DNA polymerase. In this second strand synthesis the

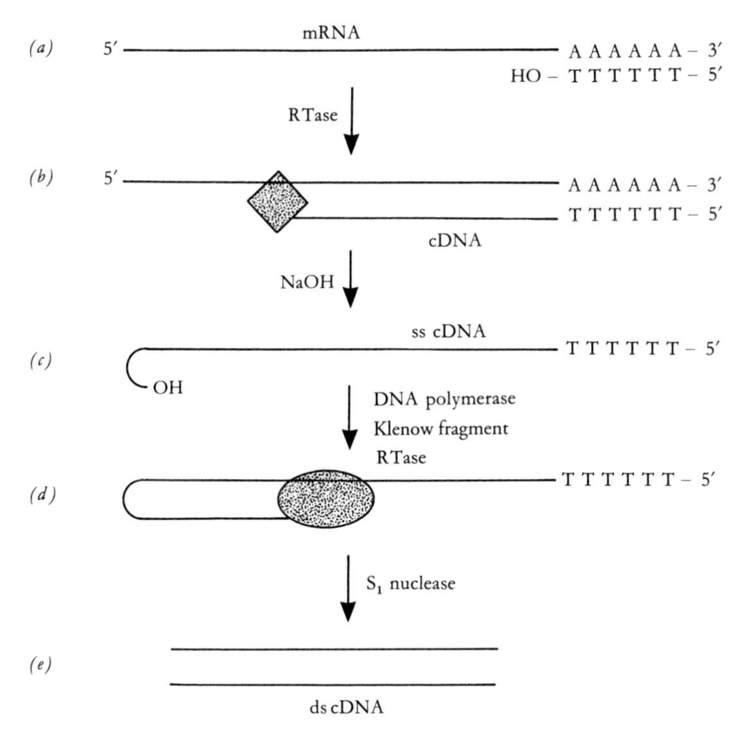

Fig. 6.2. Synthesis of cDNA. Poly(A)$^+$ RNA (mRNA) is used as the starting material. (*a*) A short oligo(dT) primer is annealed to the poly(A) tail on the mRNA, which provides the 3′-OH group for reverse transcriptase to begin copying the mRNA (*b*). The mRNA is removed by alkaline hydrolysis to give a single-stranded cDNA molecule (*c*). This has a short double-stranded hairpin loop structure which provides a 3′-OH terminus for (*d*) second strand synthesis by a DNA polymerase (T4 DNA polymerase, Klenow fragment, or RTase). (*e*) The double-stranded cDNA is trimmed with S_1 nuclease to produce a blunt-ended ds cDNA molecule. An alternative to the alkaline hydrolysis step is to use RNase H, which creates nicks in the mRNA strand of the mRNA·cDNA hybrid. By using this in conjunction with DNA polymerase I, a nick translation reaction synthesises the second cDNA strand.

priming 3′-OH is generated by short hairpin loop regions that form at the end of the ss cDNA. After second strand synthesis, the ds cDNA can be trimmed by S_1 nuclease to give a flush-ended molecule, which can then be cloned in a suitable vector.

Several problems are often encountered in synthesising cDNA using the method outlined above. Firstly, synthesis of full-length cDNAs may be inefficient, particularly if the mRNA is relatively long. This is a serious problem if expression of the cDNA is required, as it may not contain all the coding sequence of the gene. Such inefficient full-length cDNA synthesis

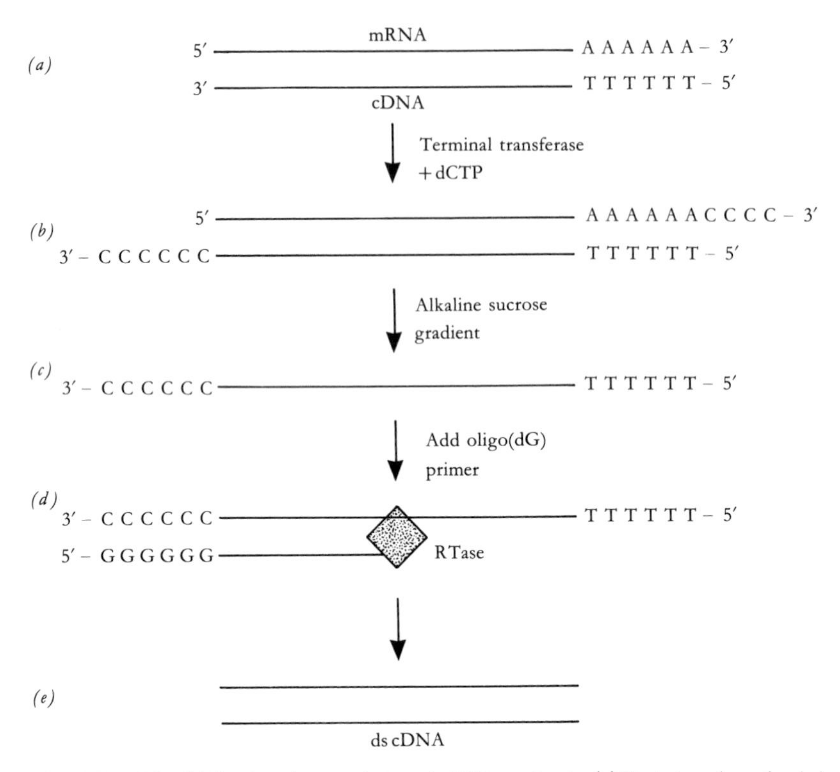

Fig. 6.3. Oligo(dG)-primed second-strand cDNA synthesis. (*a*) First strand synthesis is as shown in Fig. 6.2, generating an mRNA·cDNA hybrid. (*b*) This is tailed with C residues using terminal transferase. (*c*) Fractionation through an alkaline sucrose gradient hydrolyses the mRNA and permits recovery of full-length cDNA molecules. (*d*) An oligo(dG) primer is annealed to the C tails, and reverse transcriptase used to synthesise the second strand. (*e*) This generates a double-stranded full-length cDNA molecule. From Old and Primrose (1989), *Principles of Gene Manipulation*, 4th edition, Blackwell. Reproduced with permission.

also means that the 3′ regions of the mRNA tend to be over-represented in the cDNA population. Secondly, problems can arise from the use of S_1 nuclease, which may remove some important 5′ sequences when it is used to trim the ds cDNA.

More recent methods for cDNA synthesis overcome the above problems to a great extent, and the original method is now rarely used. One of the simplest adaptations involves the use of oligo(dC) tailing to permit oligo(dG)-primed second-strand cDNA synthesis (Fig. 6.3). The dC tails are added to the 3′ termini of the cDNA using the enzyme terminal transferase. This functions most efficiently on accessible 3′ termini, and the tailing reaction therefore favours full-length cDNAs in which the 3′ terminus is not 'hidden' by the mRNA template. The method also obviates the need for S_1 nuclease treatment, and thus full-length cDNA production is enhanced further.

Many suppliers now produce kits for cDNA synthesis. Often these have been optimised for a particular application, and the number of steps involved is usually reduced to a minimum. In many ways the mystique that surrounded cDNA synthesis in the early days has now gone, and the techniques available make full-length cDNA synthesis a relatively straightforward business. The key to success is to obtain good quality mRNA preparations and to take great care in handling these. In particular, contamination with nucleases must be avoided.

Although the poly(A) tract of eukaryotic mRNAs is often used for priming cDNA synthesis, there may be cases where this is not appropriate. Where the mRNA is not polyadenylated, random oligonucleotide primers may be used to initiate cDNA synthesis. Or, if all or part of the amino acid sequence of the desired protein is known, a specific oligonucleotide primer can be synthesised and used to initiate cDNA synthesis. This can be of great benefit in that specific mRNAs may be copied into cDNA, which simplifies the screening procedure when the clones are obtained. An additional possibility with this approach is to use the polymerase chain reaction (PCR; see section 6.4.4) to amplify selectively the desired sequence.

Having generated the cDNA fragments, the cloning procedure can begin. Here there is a further choice to be made regarding the vector system – plasmid or phage, or perhaps cosmid or phasmid? Examples of cloning strategies based on the use of plasmid and phage vectors are given below.

6.2.2 Cloning cDNA in plasmid vectors

Although many workers prefer to clone cDNA using a bacteriophage vector system, plasmids are still often used, particularly where isolation of the desired cDNA sequence involves screening a relatively small number of clones. Joining the cDNA fragments to the vector is usually achieved by one of the three methods outlined in Fig. 6.1 for cDNA cloning, these being **blunt-end ligation**, the use of **linker molecules**, and **homopolymer tailing**. Although favoured for cDNA cloning, these methods may also be used with genomic DNA (see section 6.3). Each of the three methods will be described briefly.

Blunt-end ligation is exactly what it says – the joining of DNA molecules with blunt ends, using DNA ligase (see section 4.3). In cDNA cloning, the blunt ends may arise as a consequence of the use of S_1 nuclease, or they may be generated by filling in the protruding ends with DNA polymerase. The main disadvantage of blunt-end ligation is that it is an inefficient process, as there is no specific intermolecular association to hold the DNA strands together whilst DNA ligase generates the phosphodiester linkages required to produce the recombinant DNA. Thus high concentrations of the participating DNAs must be used, so that the chances of two ends coming together are increased. The effective concentration of DNA molecules in cloning reactions is usually expressed as the concentration of termini, thus one talks about 'picomoles of ends', which can seem rather strange terminology to the uninitiated.

The conditions for ligation of ends must be chosen carefully. In theory, when vector DNA and cDNA are mixed, there are a number of possible outcomes. The desired result is for one cDNA molecule to join with one vector molecule, thus generating a recombinant with one insert. However, if concentrations are not optimal, the insert or vector DNAs may self-ligate to produce circular molecules, or the insert/vector DNAs may form concatemers instead of bimolecular recombinants. In practice, the vector is often treated with a phosphatase (either BAP or CIP; see section 4.2.3) to prevent self-ligation, and the concentrations of the vector and insert DNAs are chosen to favour the production of recombinants.

One potential disadvantage of blunt-end ligation is that it may not generate restriction enzyme recognition sequences at the cloning site, thus hampering excision of the insert from the recombinant. This is usually not a major problem, as many vectors now have a series of restriction sites clustered around the cloning site. Thus DNA inserted by blunt-end ligation can often be excised by using one of the restriction sites in the cluster. Another approach involves the use of **linkers**, which are self-complementary oligomers that contain a recognition sequence for a particular restriction enzyme. One such sequence would be 5′-CCGAATTCGG-3′, which in double-stranded form will contain the recognition sequence for *Eco*RI (GAATTC). Linkers are synthesised chemically, and can be added to cDNA by blunt-end ligation (Fig. 6.4). When they have been added, the cDNA/linker is cleaved with the linker-specific restriction enzyme, thus generating sticky ends prior to cloning. This can pose problems if the cDNA contains sites for the restriction enzyme used to cleave the linker, but these may be overcome by using a methylase to protect any internal recognition sites from digestion by the enzyme.

A second approach to cloning by addition of sequences to the ends of

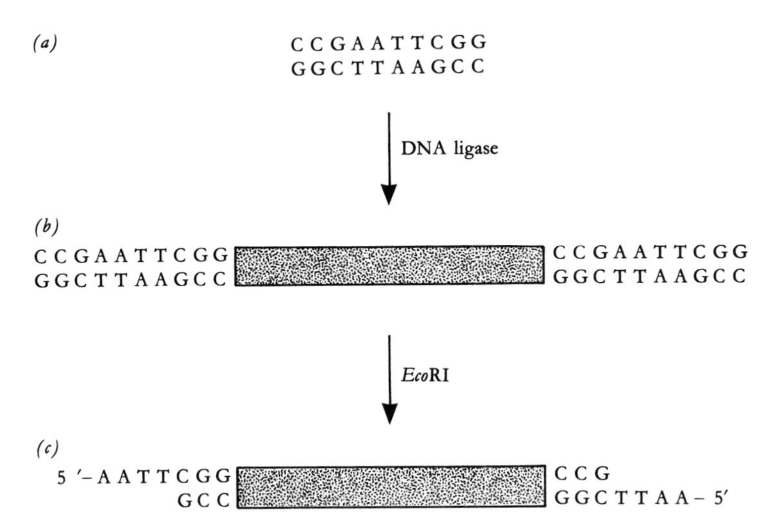

Fig. 6.4. Use of linkers. (*a*) The 10-mer 5′-CCGAATTCGG-3′ contains the recognition site for *Eco*RI. (*b*) The linker is added to blunt-ended DNA using DNA ligase. (*c*) The construct is then digested with *Eco*RI, which cleaves the linker to generate protruding 5′ termini. Redrawn from Winnacker (1987), *From Genes to Clones*, VCH. Reproduced with permission.

DNA molecules involves the use of **adaptors** (Fig. 6.5). These are single-stranded non-complementary oligomers that may be used in conjunction with linkers. When annealed together, a linker/adaptor with one blunt end and one sticky end is produced, which can be added to the cDNA to provide sticky-end cloning without digestion of the linkers.

The use of **homopolymer tailing** has proved to be a popular and effective means of cloning cDNA. In this technique, the enzyme terminal transferase (see section 4.2.3) is used to add homopolymers of dA, dT, dG or dC to a DNA molecule. Early experiments in recombinant production used dA tails on one molecule and dT tails on the other, although the technique is now most often used to clone cDNA into the *Pst*I site of a plasmid vector by dG·dC tailing. Homopolymers have two main advantages over other methods of joining DNAs from different sources. Firstly, they provide longer regions for **annealing** DNAs together than, for example, cohesive termini produced by restriction enzyme digestion. This means that **ligation** need not be carried out *in vitro*, as the cDNA·vector hybrid is stable enough to survive introduction into the host cell, where it is ligated *in vivo*. A second advantage is specificity. As the vector and insert cDNAs have different but complementary 'tails', there is little chance of

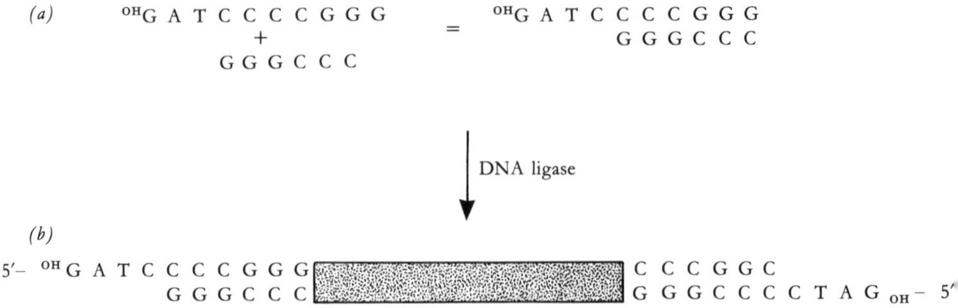

Fig. 6.5. Use of adaptors. In this example a *Bam*HI adaptor (5′-GATCCCCGGG-3′) is annealed with a single-stranded *Hpa*II linker (3′-GGGCCC-5′) to generate a double-stranded sticky-ended molecule, as shown in (*a*). This is added to blunt-ended DNA using DNA ligase. The DNA therefore gains protruding 5′ termini without the need for digestion with a restriction enzyme, as shown in (b). The 5′ terminus of the adaptor can be dephosphorylated to prevent self ligation. Redrawn from Winnacker (1987), *From Genes to Clones*, VCH. Reproduced with permission.

self-annealing, and the generation of bimolecular recombinants is favoured over a wider range of effective concentrations than is the case for other annealing/ligation reactions.

An example of the use of homopolymer tailing is shown in Fig. 6.6. The vector is cut with *Pst*I and tailed by terminal transferase in the presence of dGTP. This produces dG tails. The insert DNA is tailed with dC in a similar way, and the two can then be annealed. This regenerates the original *Pst*I site, which enables the insert to be cut out of the recombinant using this enzyme.

Introduction of cDNA·plasmid recombinants into suitable *E. coli* hosts is achieved by transformation (section 5.5.1), and the desired transformants can then be selected by the various methods available (see Chapter 7).

6.2.3 Cloning cDNA in bacteriophage vectors

Although plasmid vectors have been used extensively in cDNA cloning protocols, there are situations where they may not be appropriate. If a large number of recombinants is required, as might be the case if a low-abundance mRNA was to be cloned, phage vectors may be more suitable. The chief advantage here is that packaging *in vitro* may be used to generate the recombinant phage, which greatly increases the efficiency of the cloning process. In addition, it is much easier to store and handle large numbers of

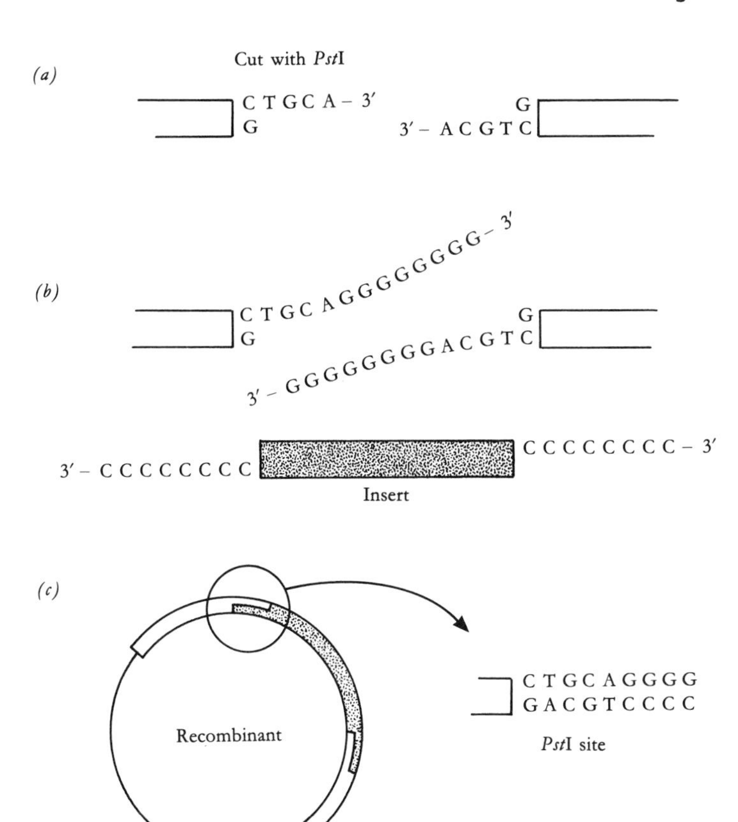

Fig. 6.6. Homopolymer tailing. (*a*) The vector is cut with *Pst*I, which generates protruding 3′-OH termini. (*b*) The vector is then tailed with dG residues using terminal transferase. The insert DNA is tailed with dC residues in a similar way. (*c*) The dC and dG tails are complementary and the insert can therefore be annealed with the vector to generate a recombinant. The *Pst*I sites are regenerated at the ends of the insert DNA, as shown.

phage clones than is the case for bacterial colonies carrying plasmids. Given that isolation of a cDNA clone of a rare mRNA species may require screening hundreds of thousands of independent clones, ease of handling becomes a major consideration.

Cloning cDNA in phage λ vectors is, in principle, no different to cloning any other piece of DNA. However, the vector has to be chosen carefully, as cDNA cloning has slightly different requirements than genomic DNA cloning in λ vectors (see section 6.3). Generally cDNAs will be much shorter than genomic DNA fragments, so an insertion vector is usually

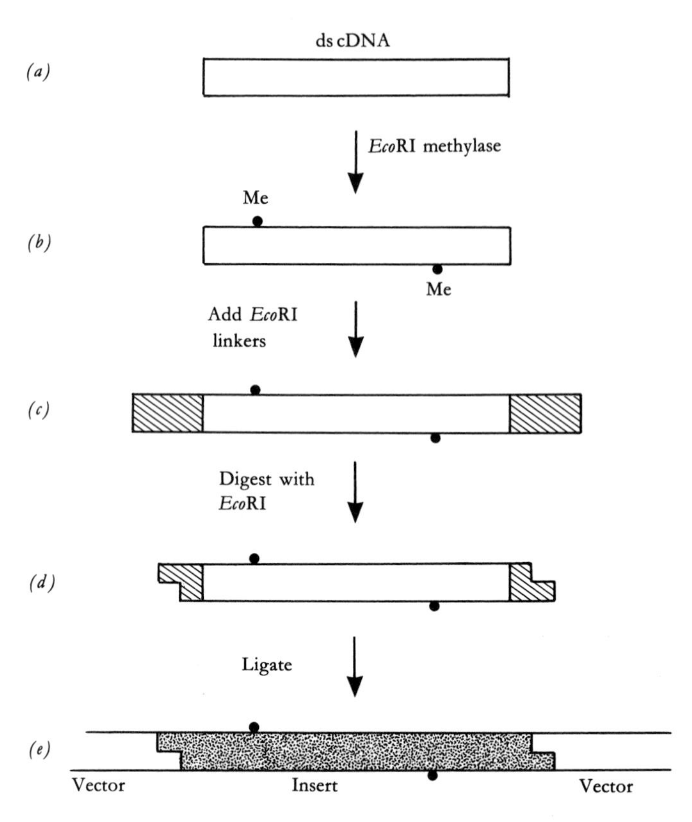

Fig. 6.7. Cloning cDNA in λ vectors using linkers. (*a*) The ds cDNA is treated with *Eco*RI methylase, which (*b*) methylates any internal *Eco*RI recognition sequences (indicated by Me). (*c*) *Eco*RI linkers are then added to the ends of the methylated cDNA, and the linkers digested with *Eco*RI. (*d*) The methylation prevents digestion at internal sites, and the result is a cDNA with *Eco*RI cohesive ends. (*e*) This can be ligated into the *Eco*RI site of a λ vector such as λgt10.

chosen. Vectors such as λgt10 and Charon 16A (section 5.3.2) are suitable, with cloning capacities of some 7.6 and 9.0 kb, respectively. The cDNA may be size-fractionated prior to cloning, to remove short cDNAs that may not be representative full-length copies of the mRNA. In the case of vectors such as λgt10, cDNA is usually ligated into the *Eco*RI site using linkers, as shown in Fig. 6.7. The recombinant DNA is packaged *in vitro* and plated on a suitable host for selection and screening.

6.3 Cloning from genomic DNA

Although cDNA cloning is an extremely useful branch of gene manipulation technology, there are certain situations where cDNAs will not provide the answers to the questions that are being posed. If, for example, the overall structure of a particular **gene** is being investigated (as opposed to its RNA transcript), the investigator may wish to determine if there are introns present. He or she will probably also wish to examine the control sequences responsible for regulating gene expression, and these will not be present in the processed mRNA molecule that is represented by a cDNA clone. In such a situation clones generated from genomic DNA must be isolated. This presents a slightly different set of problems than those involved in cloning cDNA, and therefore requires a different cloning strategy.

6.3.1 Genomic libraries

Cloning DNA, by whatever method, gives rise to a population of recombinant DNA molecules, often in plasmid or phage vectors, maintained either in bacterial cells or as phage particles. A collection of independent clones is termed a **clone bank** or **library**. The term **genomic library** is often used to describe a set of clones representing the entire genome of an organism, and the production of such a library is usually the first step in isolating a DNA sequence from an organism's genome.

What are the characteristics of a good genomic library? In theory, a genomic library should represent the entire genome of an organism as a set of overlapping cloned fragments, produced in a random manner, and maintained in a stable form with no misrepresentation of sequences. The systems available for producing genomic libraries essentially fulfil these requirements, although some compromise may be necessary during the cloning process.

The first consideration in constructing a genomic library is the number of clones required. This depends on a variety of factors, the most obvious one being the size of the genome. Thus a small genome such as that of *E. coli* will require many fewer clones than a more complex one such as the human genome. The type of vector to be used also has to be considered, which will determine size of fragments that can be cloned. In practice, library size can be calculated quite simply on the basis of the probability of a particular sequence being represented in the library. There is a formula that takes

Table 6.2. *Genomic library sizes for various organisms*

Organism	Genome size (kb)	No. clones N, P=0.95	
		20 kb inserts	45 kb inserts
Escherichia coli (bacterium)	4.0×10^3	6.0×10^2	2.7×10^2
Saccharomyces cerevisiae (yeast)	1.4×10^4	2.1×10^3	9.3×10^2
Arabidopsis thaliana (simple higher plant)	7.0×10^4	1.1×10^4	4.7×10^3
Drosophila melanogaster (fruit fly)	1.7×10^5	2.5×10^4	1.1×10^4
Stronglyocentrotus purpuratus (sea urchin)	8.6×10^5	1.3×10^5	5.7×10^4
Homo sapiens (human)	3.0×10^6	4.5×10^5	2.0×10^5
Triticum aestivum (hexaploid wheat)	1.7×10^7	2.5×10^6	1.1×10^6

Note: The number of clones (N) required for a probability (P) of 95% that a given sequence is represented is shown for various organisms. The genome sizes of the organisms are given (haploid genome size, if appropriate). Two values of N are shown, for 20 kb inserts (λ replacement vector size) and 45 kb inserts (cosmid vectors). The values should be considered as minimum estimates, as strictly speaking the calculation assumes: (i) that the genome size is known accurately, (ii) that the DNA is fragmented in a totally random manner for cloning, (iii) that each recombinant DNA molecule will give rise to a single clone, (iv) that the efficiency of cloning is the same for all fragments, and (v) that diploid organisms are homozygous for all loci. These assumptions are usually not all valid for a given experiment.

account of all the factors and produces a 'number of clones' value. The formula is:

$$N = \ln(1 - P)/\ln(1 - a/b)$$

where N is the number of clones required, P is the desired probability of a particular sequence being represented (typically set at 0.95 or 0.99), a is the average size of the DNA fragments to be cloned and b is the size of the genome (expressed in the same units as a).

By using this formula, it is possible to determine the magnitude of the task ahead, and to plan a cloning strategy accordingly. Some genome sizes and their associated library sizes are shown in Table 6.2. These library sizes should be considered as minimum values, as the generation of cloned fragments may not provide a completely random and representative set of clones in the library. Thus, for a human genomic library, we are talking of some 10^6 clones or more in order to be reasonably sure of isolating a particular single-copy gene sequence.

When dealing with this size of library, phage or cosmid vectors are usually essential, as the cloning capacity and efficiency of these vectors is much greater than that of plasmid vectors. Although cosmids, with the potential to clone fragments of up to 47 kb, would seem to be the better choice, λ replacement vectors are often used for library construction. This is because they are easier to use than cosmid vectors, and this outweighs the disadvantage of having only half the cloning capacity. In addition, the techniques for screening phage libraries are now routine and have been well characterised. This is an important consideration, particularly where workers new to the technology wish to use gene manipulation in their research.

6.3.2 Preparation of DNA fragments for cloning

One of the most important aspects of library production is the generation of genomic DNA fragments for cloning. If a λ replacement vector such as EMBL4 is to be used, the maximum cloning capacity will be around 22 kb. Thus fragments of this size must be available for the production of recombinants. In practice a range of fragment sizes is used, often between 17 and 23 kb for a vector such as EMBL4. It is important that smaller fragments are not ligated into the vector, as there is the possibility of multiple inserts which could arise by ligation of small non-contiguous DNA sequences into the vector.

There are two main considerations when preparing DNA fragments for cloning, these being: (i) the molecular weight of the DNA after isolation from the organism, and (ii) the method used to fragment the DNA. For a completely random library, the starting material should be very high molecular weight DNA, and this should be fragmented by a totally random (i.e. sequence-independent) method. Isolation of DNA in excess of 100 kb in length is desirable, and this in itself can pose technical difficulties where the type of source tissue does not permit gentle disruption of cells. In addition, pipetting and mixing solutions of high molecular weight DNA can cause shearing of the molecules, and great care must be taken when handling the preparations.

Assuming that sufficient DNA of 100 kb is available, fragmentation can be carried out. This is usually followed by a size-selection procedure to isolate fragments in the desired range of sizes. Fragmentation can be achieved either by mechanical shearing or by partial digestion with a restriction enzyme. Although mechanical shearing (by forcing the DNA through a syringe needle, or by sonication) will generate random fragments, it will not produce DNA with cohesive termini. Thus further manipulation such as trimming or filling in the ragged ends of the molecules will be required before the DNA can be joined to the vector, usually with linkers, adaptors or homopolymer tails (see Fig. 6.1). In practice these additional steps are often considered undesirable, and fragmentation by partial restriction digestion is used extensively in library construction. However, this is not a totally sequence-independent process, as the occurrence of restriction enzyme recognition sites is clearly sequence-dependent. Partial digestion is therefore something of a compromise, but careful design and implementation of the procedure can overcome most of the disadvantages.

If a restriction enzyme is used to digest DNA to completion, the fragment pattern will obviously depend on the precise location of recognition sequences. This approach therefore has two drawbacks. Firstly, a six-cutter such as *Eco*RI will have recognition sites on average about once every 4096 bp, which would produce fragments that are too short for λ replacement vectors. Secondly, any sequence bias, perhaps in the form of repetitive sequences, may skew the distribution of recognition sites for a particular enzyme. Thus some areas of the genome may contain few sites, whilst others have an over-abundance. This means that a complete digest will not be suitable for generating a representative library. If however a **partial** digest is carried out using an enzyme that cuts frequently (e.g. a four-cutter such as *Sau*3A, which cuts on average once every 256 bp), the effect is to produce a collection of fragments that are essentially random. This can be achieved by varying the enzyme concentration or the time of digestion, and a test run will produce a set of digests which contain different fragment size distribution profiles, as shown in Fig. 6.8.

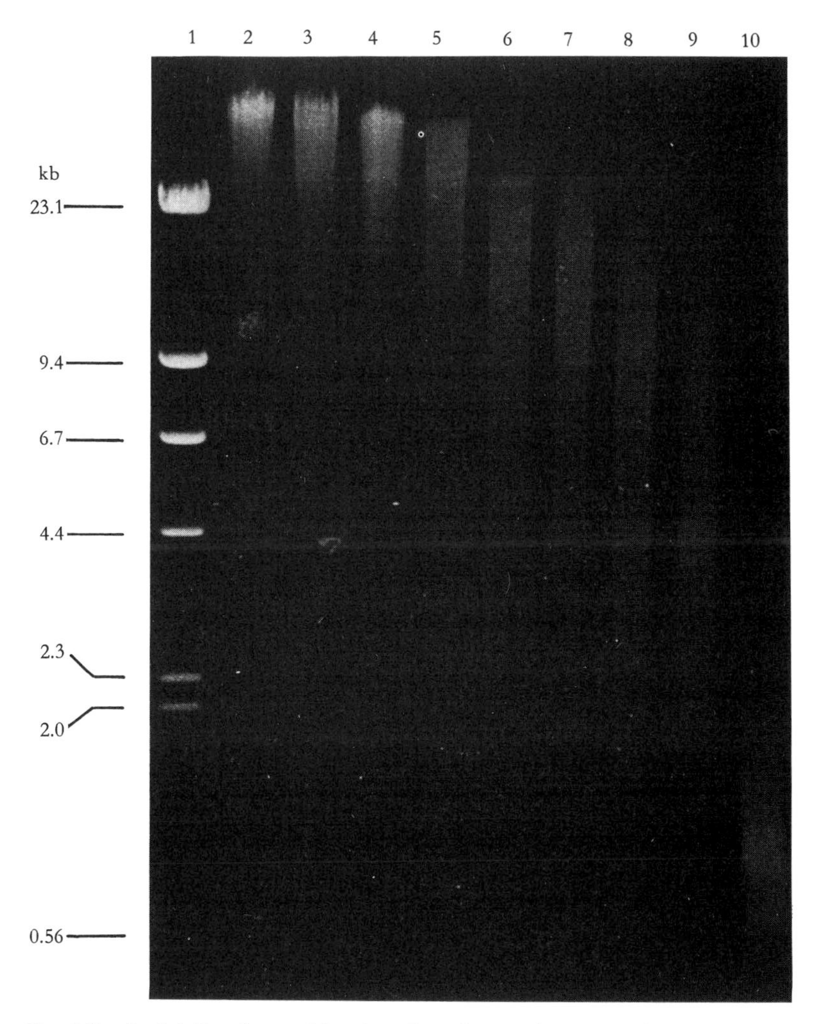

Fig. 6.8. Partial digestion and fractionation of genomic DNA. High molecular weight genomic DNA was digested with various concentrations of the restriction enzyme *Sau*3A. Samples from each digest were run on a 0.7% agarose gel and stained with ethidium bromide. Lane 1 shows λ *Hin*dIII markers, sizes as indicated. Lanes 2 to 10 show the effects of increasing concentrations of restriction enzyme in the digestions. As the concentration of enzyme is increased, the DNA fragments generated are smaller. From this information the optimum concentration of enzyme to produce fragments of a certain size distribution can be determined. These can then be run on a gel (as here) and isolated prior to cloning. Photograph courtesy of Dr N. Urwin.

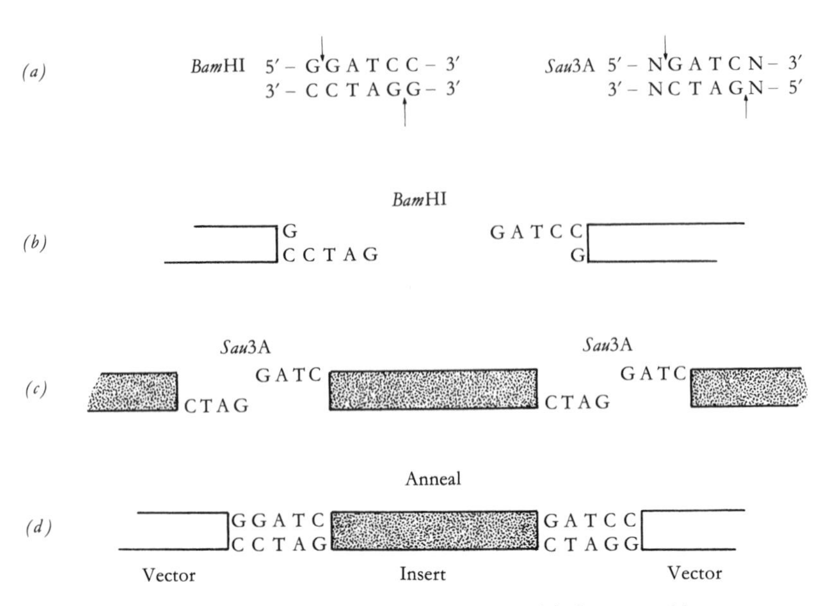

Fig. 6.9. Cloning *Sau*3A fragments into a *Bam*HI site. (*a*) The recognition sequences and cutting sites for *Bam*HI and *Sau*3A. In the *Sau* 3A site, N is any base. (*b*) Vector DNA cut with *Bam*HI generates 5′ protruding termini with the four-base sequence 5′-GATC-3′. (*c*) Insert DNA cut with *Sau*3A also generates identical four-base overhangs. (*d*) Thus DNA cut with *Sau*3A can be annealed to *Bam*HI cohesive ends to generate a recombinant DNA molecule.

6.3.3 Ligation, packaging and amplification of libraries

Having established the optimum conditions for partial digestion, a sample of DNA can be prepared for cloning. After digestion the sample is fractionated, either by density gradient centrifugation or by electrophoresis. Fragments in the range 17–23 kb can then be selected for ligation. If *Sau*3A (or *Mbo*I, which has the same recognition sequence) has been used as the digesting enzyme, the fragments can be inserted into the *Bam*HI site of a vector such as EMBL4, as the ends generated by these enzymes are complementary (Fig. 6.9). The insert DNA can be treated with phosphatase to reduce self-ligation or concatemer formation, and the vector can be digested with *Bam*HI and *Sal*I to generate the cohesive ends for cloning and to isolate the stuffer fragment and prevent it from re-annealing during ligation. The *Eco*RI site in the vector can be used to excise the insert after cloning. Ligation of DNA into EMBL4 is summarised in Fig. 6.10.

When ligation is carried out, concatemeric recombinant DNA molecules

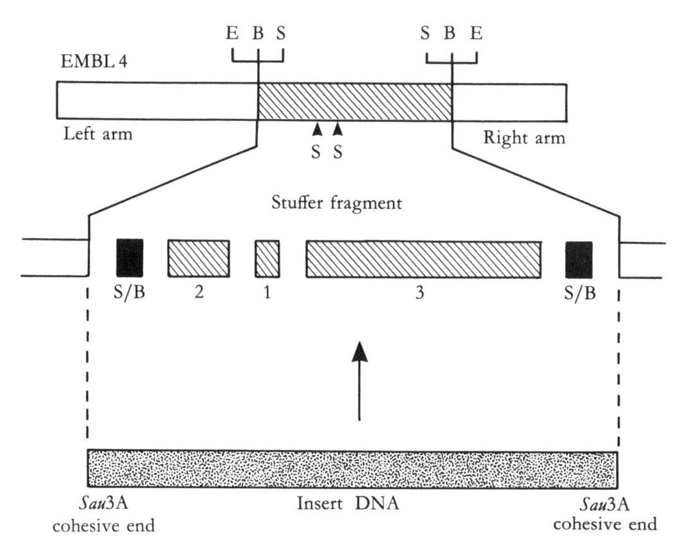

Fig. 6.10. Ligation of *Sau*3A-cut DNA into the λ replacement vector EMBL4. Sites on the vector are *Eco*RI (E), *Bam*HI (B) and *Sal*I (S). The vector is cut with *Bam*HI and *Sal*I, which generates five fragments from the stuffer fragment (hatched in top panel). Removal of the very short *Sal*I/*Bam*HI fragments (filled boxes; not drawn to scale) prevents the stuffer fragment from re-annealing. In addition, the two internal *Sal*I sites cleave the stuffer fragment, producing three *Sal*I/*Sal*I fragments (1 to 3). If desired, the short fragments can be removed from the preparation by precipitation with isopropanol, which leaves the small fragments in the supernatant. On removal of the stuffer, *Sau*3A-digested insert DNA can be ligated into the *Bam*HI site of the vector (see Fig. 6.9).

are produced, which are suitable substrates for packaging *in vitro*, as shown in Fig. 6.11. This produces what is known as a **primary library**, which consists of individual recombinant phage particles. Whilst this is theoretically the most useful type of library in terms of isolation of a specific sequence, it is a finite resource. Thus a primary library is produced, screened and then discarded. If the sequence of interest has not been isolated, more recombinant DNA will have to be produced and packaged. Whilst this may not be a problem, there are occasions where a library may be screened for several different genes, or may be sent to different laboratories. In these cases it is therefore necessary to **amplify** the library. This is achieved by plating the packaged phage on a suitable host strain of *E. coli*, and then resuspending the plaques by gently washing the plates with a buffer solution. The resulting phage suspension can be stored almost indefinitely, and will provide enough material for many screening and isolation procedures.

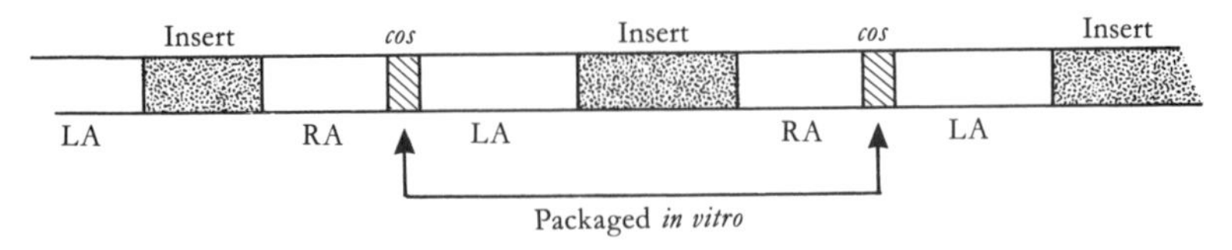

Fig. 6.11. Concatemeric recombinant DNA. On ligation of inserts into a vector such as EMBL4, a concatemer is formed. This consists of the left arm of the vector (LA), the insert DNA, and the right arm (RA). These components of the unit are repeated many times and are linked together at the *cos* sites by the cohesive ends on the vector arms. On packaging *in vitro*, the recombinant genomes are cut at the *cos* sites and packaged into phage heads.

Although amplification is a useful step in producing stable libraries, it can lead to skewing of the library. Some recombinant phage may be lost, perhaps due to the presence of repetitive sequences in the insert, which can give rise to recombinational instability. This can be minimised by plating on a recombination-deficient host strain. Some phage may exhibit differential growth characteristics which may cause particular phage to be either over- or under-represented in the amplified library, and this may mean that a greater number of plaques have to be screened in order to isolate the desired sequence.

6.4 Advanced cloning strategies

In sections 6.2 and 6.3 I examined cDNA and genomic DNA cloning strategies, using basic plasmid and phage vectors in *E. coli* hosts. These approaches have proved to be both reliable and widely applicable, and represent a major part of the technology of gene manipulation. However, advances made over the past few years have increased the scope (and often the complexity!) of cloning procedures. Such advances include more sophisticated vectors for *E. coli* and other hosts, increased use of expression vectors, and novel approaches to various technical problems. Some examples of more advanced cloning strategies are discussed below.

6.4.1 Synthesis and cloning of cDNA

An elegant scheme for generating cDNA clones was developed by Hiroto Okayama and Paul Berg in 1982. In their method the plasmid vector itself is used as the **priming** molecule, and the mRNA is annealed to this for cDNA

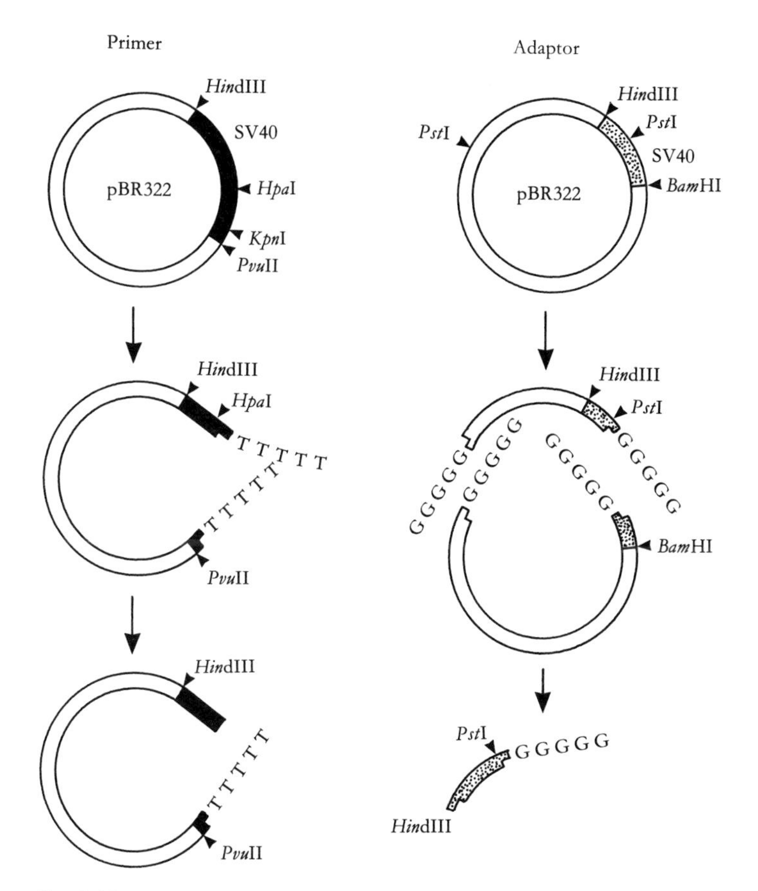

Fig. 6.12. Preparation of vector and adaptor molecules for Okayama and Berg cDNA cloning. The vector is made up from pBR322 plus parts of the SV40 genome (solid or shaded in the diagram). For the primer, the vector is cut with *Kpn*I and tailed with dT residues. It is then digested with *Hpa*I to create a vector in which one end is tailed. The adaptor molecule is generated by cutting the adaptor plasmid with *Pst*I, which generates two fragments. These are tailed with dG residues and digested with *Hin*dIII to produce the adaptor molecule itself, which therefore has a *Hin*dIII cohesive end in addition to the dG tail. The fragment is purified for use in the protocol (see Fig. 6.13). From Old and Primrose (1989), *Principles of Gene Manipulation*, 4th edition, Blackwell. Reproduced with permission.

synthesis. A second **adaptor** molecule is required to complete the process. Both adaptor and primer are based on pBR322, with additional sequences from the SV40 virus. Preparation of the vector and adaptor molecules involves restriction digestion, tailing with oligo(dT) and dG, and purification of the fragments to give the molecules shown in Fig. 6.12. The mRNA

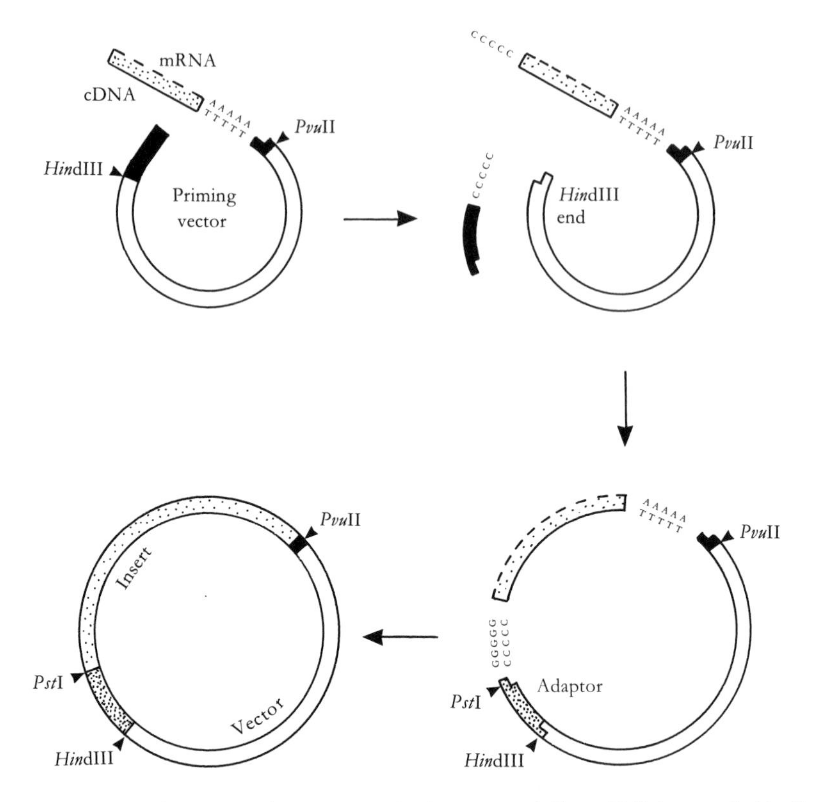

Fig. 6.13. Okayama and Berg cDNA cloning protocol. The mRNA is annealed to the dT-tailed priming vector *via* the poly(A) tail, and the first cDNA strand synthesised using reverse transcriptase. The cDNA is then tailed with dC residues and the dC-tailed vector fragment removed by digestion with *Hin*dIII. The cDNA is annealed to the dG tailed adaptor molecule, which is ligated into the vector using the cohesive *Hin*dIII ends on the vector and adaptor molecules. Finally the mRNA is displaced and the second cDNA strand synthesised using RNase H and DNA polymerase I to generate the complete vector/insert recombinant. From Old and Primrose (1989), *Principles of Gene Manipulation*, 4th edition, Blackwell. Reproduced with permission.

is then annealed to the plasmid and the first cDNA strand synthesised and tailed with dC. The terminal vector fragment (which is also tailed during this procedure) is removed and the adaptor added to circularise the vector prior to synthesis of the second strand of the cDNA. Second-strand synthesis involves the use of RNase H, DNA polymerase I and DNA ligase in a strand-replacement reaction which converts the mRNA·cDNA hybrid into ds cDNA and completes the ligation of the ds cDNA into the vector. The end result is that recombinants are generated in which there is a high proportion of full-length cDNAs. The Okayama and Berg method is summarised in Fig. 6.13.

6.4.2 Expression of cloned cDNA molecules

Many of the routine manipulations in gene cloning experiments do not require expression of the cloned DNA. However, there are certain situations in which some degree of genetic expression is needed. The obvious example is where the recombinant DNA is used to produce a protein, which involves both transcription and translation of the cloned sequence. If eukaryotic DNA sequences are cloned, post-transcriptional and post-translational modifications may be required, and the type of host/vector system that is used is therefore very important in determining whether or not such sequences will be expressed effectively. The problem of RNA processing in prokaryotic host organisms may be obviated by cloning cDNA sequences, and this is the most common approach where expression of eukaryotic sequences is desired. In this section I consider some aspects of cloning cDNAs for expression, concentrating mainly on the characteristics of the vector/insert combination that enable expression to be achieved. Further discussion of the topic is presented in section 8.2.

Assuming that a functional cDNA sequence is available, a suitable host/vector combination must be chosen. The host cell type will usually have been selected by considering aspects such as ease of use, fermentation characteristics or the ability to secrete proteins derived from cloned DNA. However, for a given host cell, there may be several types of expression vector, including both plasmid and (for bacteria) phage-based examples. In addition to the normal requirements such as restriction site availability and genetic selection mechanisms, a key feature of expression vectors is the type of **promoter** that is used to direct expression of the cloned sequence. Often the aim will be to maximise the expression of the cloned sequence, so a vector with a highly efficient promoter is chosen. Such promoters are often termed **strong** promoters. However, if the product of the cloned gene is toxic to the cell, a **weak** promoter may be required to avoid cell death due to over-expression of the toxic product.

Promoters are regions with a specific base sequence, to which RNA polymerase will bind. By examining the base sequence lying on the 5′ (upstream) side of the coding regions of many different genes, the types of sequences that are important have been identified. Although there are variations, these sequences all have some similarities. The 'best fit' sequence for a region such as a promoter is known as the **consensus sequence**. In prokaryotes there are two main regions that are important. Some 10 base-pairs upstream from the transcription start site (the −10 region, as the T_C start site is numbered +1) there is a region known as the **Pribnow box**, which has the consensus sequence 5′-TATAAT-3′. A second important region is located around position −35, and has the consensus sequence 5′-TTGACA-3′. These two regions form the basis of promoter structures in prokaryotic cells, with the precise sequences found in each region determining the strength of the promoter.

Sequences important for transcription initiation in eukaryotes have been identified in much the same way as for prokaryotes. Eukaryotic promoter structure is generally more complex than that found in prokaryotes, and control of initiation of transcription can involve sequences (e.g. enhancers) that may be several hundred or thousand base-pairs upstream from the T_C start site. However, there are important motifs closer to the start site. These are a region centred around position −25 with the consensus sequence 5′-TATAAAT-3′ (the **TATA** or **Hogness** box) and a sequence in the −75 region with the consensus 5′-GG(T/C)CAATCT-3′, known as the **CAAT** box.

In addition to the strength of the promoter, it may be desirable to regulate the expression of the cloned cDNA by using promoters from genes that are either **inducible** or **repressible**. Thus some degree of control can be exerted over the transcriptional activity of the promoter; when the cDNA product is required, transcription can be 'switched on' by manipulating the system using an appropriate metabolite. Some examples of promoters used in the construction of expression vectors are given in Table 6.3.

In theory, constructing an expression vector is straightforward once a suitable promoter has been identified. In practice, as is often the case, the process is often highly complex, requiring many manipulations before a functional vector is obtained. The basic vector (often a plasmid) must carry an origin of replication that is functional in the target host cell, and there may be antibiotic resistance genes or other genetic selection mechanisms present. However, as far as expression of cloned sequences is concerned, it is the arrangement of restriction sites immediately downstream from the promoter that is critical. There must be a unique restriction site for cloning into, and this has to be located in a position where the inserted cDNA sequence can be expressed effectively. This aspect of vector structure is discussed further in section 8.2.

6.4.3 Cloning large DNA fragments in YAC vectors

Yeast artificial chromosomes (YACs: see Fig. 5.11) can be used to clone very long pieces of DNA. The use of YAC vectors can reduce dramatically

Table 6.3. *Promoters used in expression vectors*

Organism	Gene promoter	Induction by
E. coli	*lac* operon	IPTG
	trp operon	ß-Indolylacetic acid
	λP_L	Temperature-sensitive λ cI protein
A. nidulans	Glucoamylase	Starch
S. cerevisiae	Acid phosphatase	Phosphate depletion
	Alcohol dehydrogenase	Glucose depletion
	Galactose utilization	Galactose
	Metallothionein	Heavy metals
T. reesei	Cellobiohydrolase	Cellulose
Mouse	Metallothionein	Heavy metals
Human	Heat-shock protein	Temperature >40 °C

Note: Some examples of various promoters that can be used in expression vectors are given, with the organism from which the gene promoter is taken. The conditions under which gene expression is induced from such promoters are also given.
Source: Collated from Brown (1990), Gene Cloning, Chapman & Hall; and Old & Primrose (1989), Principles of Gene Manipulation, Blackwell. Reproduced with permission.

the number of clones needed to produce a representative genomic library for a particular organism, and this is a desirable outcome in itself. A consequence of cloning large pieces of DNA is that physical mapping of genomes is made simpler, as there are not as many non-contiguous sequences to fit together in the correct order.

A further advantage of cloning long stretches of DNA stems from the fact that many eukaryotic genes are much larger than the 47 kb or so that can be cloned using cosmid vectors in *E. coli*. Thus with plasmid, phage and cosmid vectors it may be impossible to isolate the entire gene. This makes it difficult to determine gene structure without using several different clones, which is not the ideal way to proceed. The use of YAC vectors can alleviate this problem and can enable the structure of large genes to be determined by providing a single DNA fragment to work from.

In practice, cloning in YAC vectors is similar to other protocols (Fig. 6.14). The vector is prepared by a double restriction digest, which releases the vector sequence between the telomeres and cleaves the vector at the cloning site. Thus two arms are produced, as is the case with phage

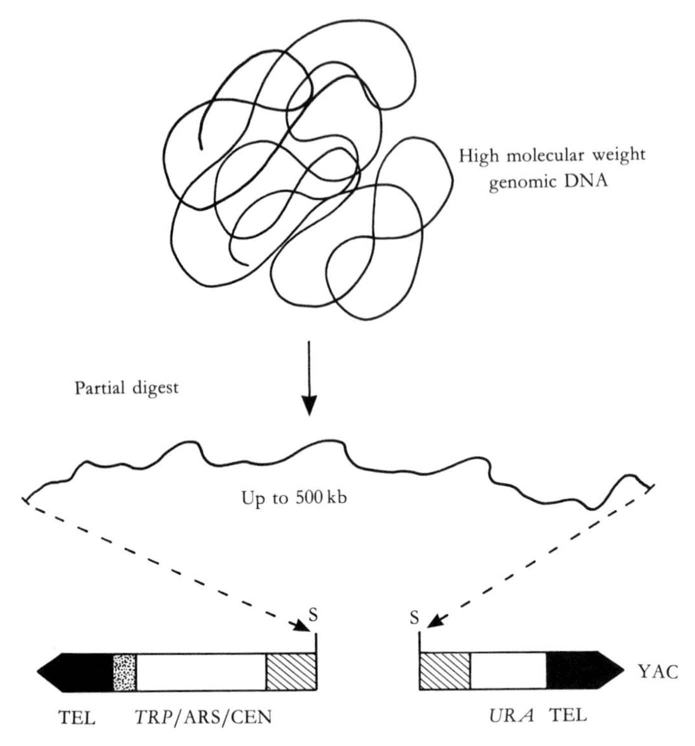

Fig. 6.14. Cloning in a YAC vector. Very large DNA fragments (up to 500 kb) are generated from high molecular weight DNA. The fragments are then ligated into a YAC vector (see Fig. 5.11) that has been cut with *Bam*HI and *Sma*I (S). The construct contains the cloned DNA and the essential requirements for a yeast chromosome i.e. telomeres (TEL), an autonomous replication sequence (ARS) and a centromere region (CEN). The *TRP* and *URA* genes can be used as dual selectable markers to ensure that only complete artificial chromosomes are maintained. From Kingsman and Kingsman (1988), *Genetic Engineering*, Blackwell. Reproduced with permission.

vectors. Insert DNA is prepared as very long fragments (a partial digest with a six-cutter may be used) and ligated into the cloning site to produce artificial chromosomes. Selectable markers on each of the two arms ensure that only correctly constructed chromosomes will be selected and propagated.

6.4.4 The polymerase chain reaction

Over the past few years many areas of molecular biology have been revolutionised by a technique which enables selective amplification of

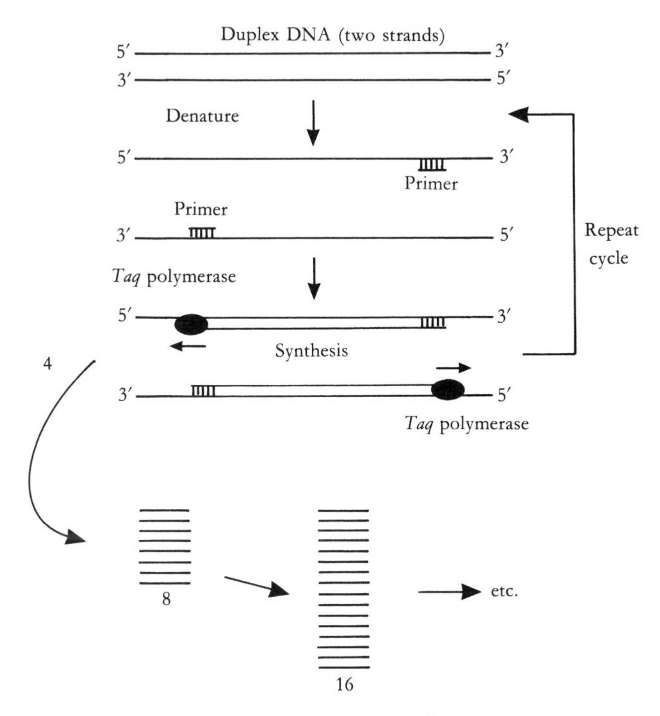

Fig. 6.15. The polymerase chain reaction (PCR). Duplex DNA is heat denatured to give single strands, and two oligonucleotide primers are annealed to their complementary sequences on the target DNA. *Taq* polymerase (thermostable) is used to synthesise complementary strands from the template strands by primer extension. The cycle is then repeated by denaturation of the DNA, and the denature/prime/copy programme repeated many times. The numbers refer to the number of DNA strands in the reaction; the duplex starts off as 2, there are 4 at the end of the first cycle, 8 at the end of the second, 16 at the end of the third, and so on. In theory, after 30 cycles 2.15×10^9 strands are produced. Thus DNA sequences can be amplified very quickly.

DNA sequences. The technique is known as the **polymerase chain reaction** (PCR), and is elegantly simple in theory (see Fig. 6.15). When a DNA duplex is heated, the strands separate or 'melt', generating single-stranded sequences. If these sequences can be copied by DNA polymerase, the original sequence is effectively duplicated. If the process is repeated many times, there is an exponential increase in the number of copies of the starting sequence. Thus, after relatively few cycles, the target sequence becomes greatly amplified, which can be invaluable in the identification and further processing of the sequence.

In addition to a DNA sequence for amplification, there are two requirements for PCR. Firstly, a suitable primer is required. In practice, *two* primers are necessary, one for each strand of the duplex. The primers should flank the target sequence, so some sequence information is required if selective amplification is to be achieved. The primers are synthesised as oligonucleotides, and are added to the reaction in excess, so that each of the primers is always available following the denaturation step. A second requirement, which makes life much easier for the operator, is the availability of a thermostable form of DNA polymerase. This is purified from the thermophilic bacterium *Thermus aquaticus*, which inhabits hot springs. The use of *Taq* polymerase means that the PCR procedure can be automated, as there is no need to add fresh polymerase after each denaturation step, as would be the case if a heat-sensitive enzyme was used.

In operation, PCR is straightforward. The target DNA is denatured by heating (usually to 80 °C or more) and the priming oligonucleotides added with the other reaction components. As the temperature drops, primers will anneal to their target sequences on the ss DNA. The reaction is started by the addition of *Taq* polymerase, and synthesis allowed to proceed. The cycle is completed (and re-started) by a further denaturation step.

Applications of PCR technology are many and diverse. It can be used to clone specific sequences, although in many cases it is in fact not necessary to do this, as enough material for subsequent manipulations may be produced by the PCR process itself. It can be used to clone genes from one organism by using priming sequences from another, if some sequence data are available for the gene in question. Another use of the PCR process is in forensic and diagnostic procedures such as the examination of bloodstains or in antenatal screening for genetic disorders. These areas are particularly important in the wider context of the applications of gene technology, and are discussed further in Chapter 8.

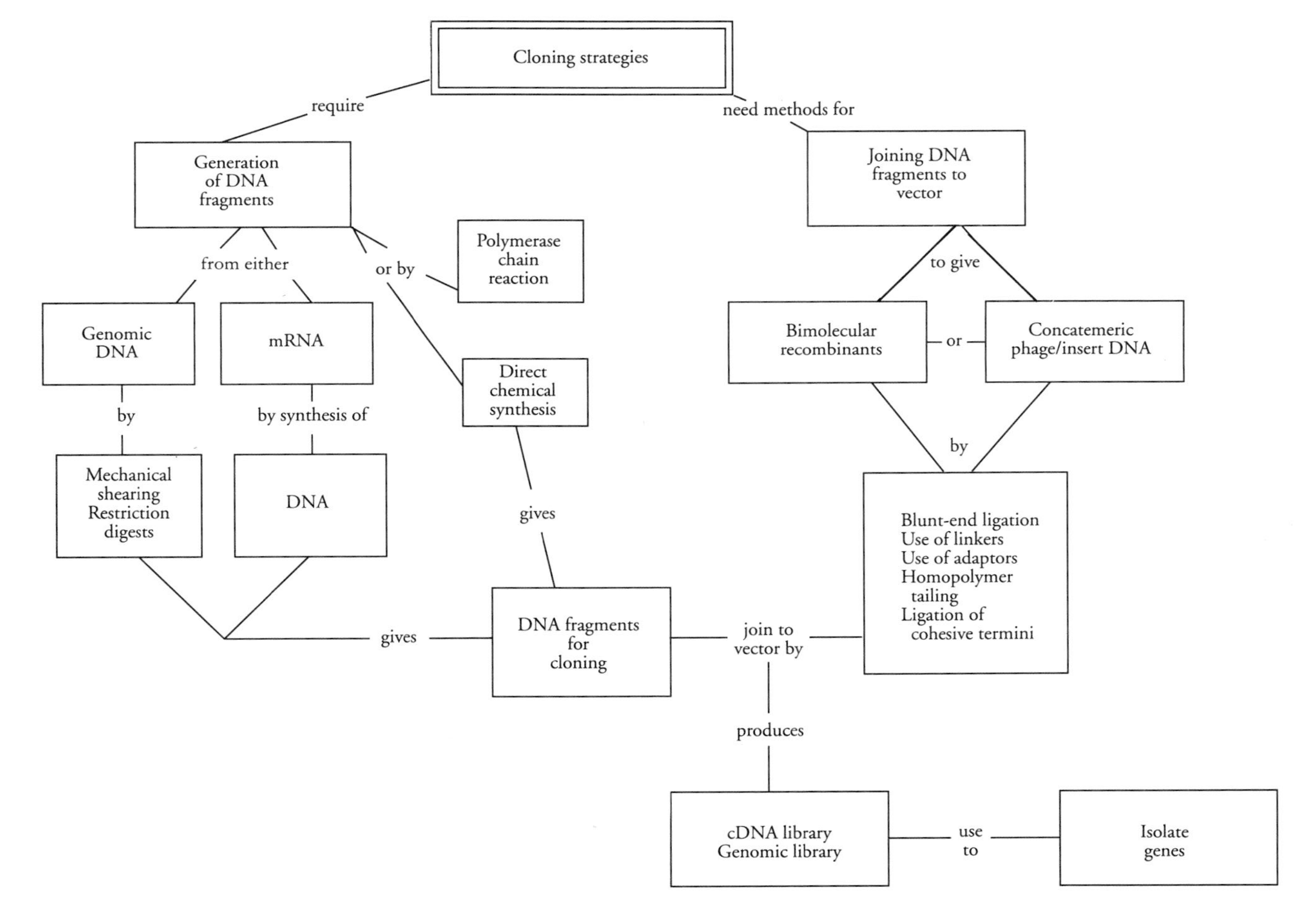

Concept map 6.

7

Selection, screening and analysis of recombinants

Success in any cloning experiment depends on being able to identify the desired gene sequence among the many different recombinants that may be produced. Given that a large genomic library may contain a million or more cloned sequences, which are not readily distinguishable from each other by simple analytical methods, it is clear that identification of the target gene is potentially the most difficult part of the cloning process. Fortunately there are several selection/identification methods that can be used to overcome most of the problems that arise. In this chapter I examine some of these methods, and also consider some of the techniques that are used to characterise cloned genes once they have been identified.

There are two terms that require definition before we proceed, these being **selection** and **screening**. Selection is where some sort of pressure (e.g. the presence of an antibiotic) is applied during the growth of host cells containing recombinant DNA. The cells with the desired characteristics are therefore **selected** by their ability to survive. This approach ranges in sophistication, from simple selection for the presence of a vector, up to direct selection of cloned genes by complementation of defined mutations. **Screening** is a procedure by which a population of viable cells is subjected to some sort of analysis that enables the desired sequences to be identified. Because only a small proportion of the large number of bacterial colonies or bacteriophage plaques being screened will contain the DNA sequence(s) of interest, screening requires methods that are highly sensitive and specific. In practice, both selection and screening methods may be required in any single experiment, and may even be used at the same time if the procedure is designed carefully.

7.1 Genetic selection and screening methods

Genetic selection and screening methods rely on the expression (or non-expression) of certain traits. Usually these traits are encoded by the vector, or perhaps by the desired cloned sequence if a direct selection method is available.

One of the simplest genetic selection methods involves the use of antibiotics to select for the presence of vector molecules. For example, the plasmid pBR322 contains genes for ampicillin resistance (Ap^r) and tetracycline resistance (Tc^r). Thus the presence of the plasmid in cells can be detected by plating potential transformants on an agar medium that contains either (or both) of these antibiotics. Only cells that have taken up the plasmid will be resistant, and these cells will therefore grow in the presence of the antibiotic.

Genetic selection methods can be simple (as above) or complex, depending on the characteristics of the vector/insert combination, and on the type of host strain used. Such methods are extremely powerful, and there is a wide variety of genetic selection and screening techniques available for many diverse applications. Some of these are described below.

7.1.1 The use of chromogenic substrates

The use of chromogenic substrates in genetic screening methods has been an important aspect of the development of the technology. The most popular system uses the compound X-gal (5-bromo-4-chloro-3-indolyl-ß-D-galactopyranoside), which is a colourless substrate for ß-galactosidase. The enzyme is normally synthesised by *E. coli* cells when lactose becomes available. However, induction can also occur if a lactose analogue such as IPTG (iso-propyl-thiogalactoside) is used. This has the advantage of being an inducer without being a substrate for ß-galactosidase. On cleavage of X-gal a blue-coloured product is formed (Fig. 7.1), thus the expression of the *lacZ* (ß-galactosidase) gene can be detected easily. This can be used either as a screening method for cells or plaques, or as a system for the detection of tissue-specific gene expression in transgenics (see section 8.4.3).

The X-gal detection system can be used where a functional ß-galactosidase gene is present in the host/vector system. This can occur in two ways. Firstly, an intact ß-galactosidase gene (*lacZ*) may be present in the vector, as is the case for the λ insertion vector Charon 16A (see Fig. 5.8). Host cells

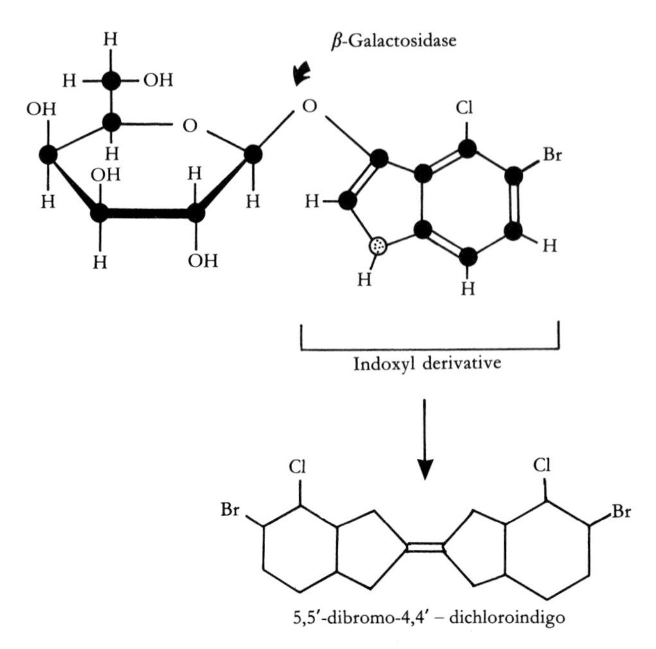

Fig. 7.1. Structure of X-gal and cleavage by ß-galactosidase. The colourless compound X-gal (5-bromo-4-chloro-3-indolyl-ß-D-galactopyranoside) is cleaved by ß-galactosidase to give galactose and an indoxyl derivative. This derivative is in turn oxidised in air to generate the dibromo–dichloro derivative, which is blue.

that are Lac$^-$ are used for propagation of the phage, so that the Lac$^+$ phenotype will only arise when the vector is present. A second approach is to employ the α-complementation system, in which part of the *lacZ* gene is carried by the vector and the remaining part is carried by the host cell. The smaller, vector-encoded peptide fragment is known as the α-peptide, and the region coding for this is designated *lacZ′*. Host cells are therefore designated *lacZ′*$^-$. Blue colonies or plaques will only be produced when the host and vector fragments complement each other to produce functional ß-galactosidase.

7.1.2 Insertional inactivation

The presence of cloned DNA fragments can be detected if the insert interrupts the coding sequence of a gene. This approach is known as **insertional inactivation**, and can be used with any suitable genetic system. Three systems will be described to illustrate the use of the technique.

Antibiotic resistance can be used as an insertional inactivation system if DNA fragments are cloned into a restriction site within an antibiotic-resistance gene. For example, cloning DNA into the *Pst*I site of pBR322 (which lies within the Ap^r gene) interrupts the coding sequence of the gene, and renders it non-functional. Thus cells that harbour a recombinant plasmid will be Ap^sTc^r. This can be used to identify recombinants as follows: if transformants are plated firstly onto a tetracycline-containing medium, all cells that contain the plasmid will survive and form colonies. If a replica of the plate is then taken and grown on ampicillin-containing medium, the recombinants (Ap^sTc^r) will not grow, but any non-recombinant transformants (Ap^rTc^r) will. Thus recombinants are identified by their absence from the replica plate, and can be picked from the original plate and used for further analysis.

The X-gal system can also be used as a screen for cloned sequences. If a DNA fragment is cloned into a functional ß- galactosidase gene (e.g. into the *Eco*RI site of Charon 16A), any recombinants will be genotypically $lacZ^-$ and will therefore not produce ß-galactosidase in the presence of IPTG and X-gal. Plaques containing such phage will therefore remain colourless. Non-recombinant phage will retain a functional *lacZ* gene, and therefore give rise to blue plaques. This approach can also be used with the α-complementation system; in this case the insert DNA inactivates the *lacZ′* region in vectors such as the M13 phage and pUC plasmid series. Thus complementation will not occur in recombinants, which will be phenotypically Lac^- and will therefore give rise to colourless plaques or colonies (Fig. 7.2).

Plaque morphology can also be used as a screening method for certain λ vectors such as λgt10, which contain the *cI* gene. This gene encodes the cI repressor, which is responsible for the formation of lysogens. Plaques derived from cI^+ vectors will be slightly turbid, due to the survival of some cells that have become lysogens. If the *cI* gene is inactivated by cloning a fragment into a restriction site within the gene, the plaques are clear and can be distinguished from the turbid non-recombinants. This system can also be used as a selection method (see section 7.1.4).

7.1.3 Complementation of defined mutations

Direct selection of cloned sequences is possible in some cases. An example is where antibiotic resistance genes are being cloned, as the presence of cloned sequences can be detected by plating cells on a medium that contains

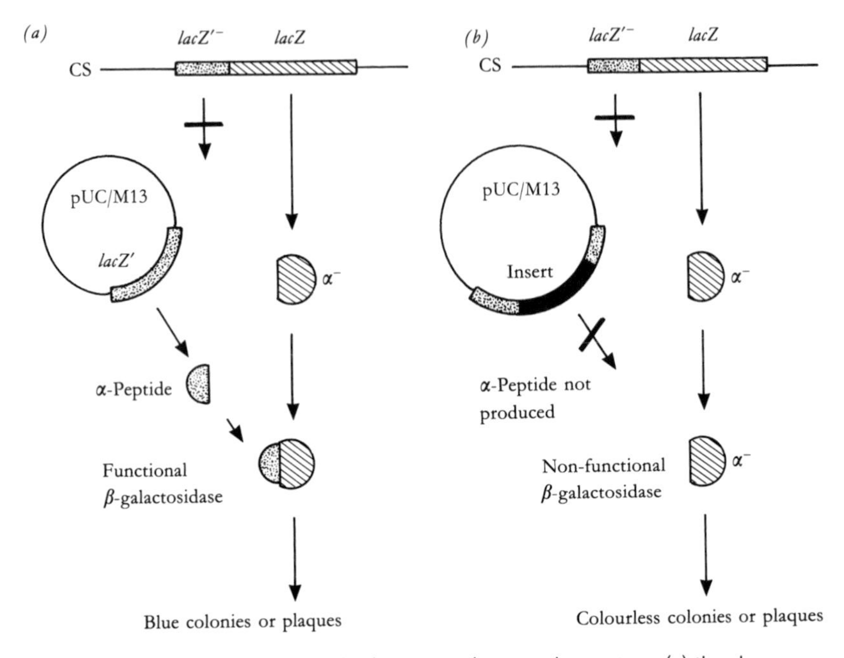

Fig. 7.2. Insertional inactivation in the α-complementation system. (*a*) the chromosome (CS) has a defective *lacZ* gene that does not encode the N-terminal α-peptide of ß-galactosidase (specified by the *lacZ*′⁻ gene fragment). Thus the product of the chromosomal *lacZ* region is an enzyme lacking the α-peptide (α^-, hatched). If a non-recombinant pUC plasmid or M13 phage is present in the cell, the *lacZ*′ gene fragment encodes the α-peptide, which enables functional ß-galactosidase to be produced. In the presence of X-gal, blue colonies or plaques will appear. If a DNA fragment is cloned into the vector, as shown in (*b*), the *lacZ*′ gene is inactivated and no complementation occurs. Thus colonies or plaques will not appear blue.

the antibiotic in question (assuming that the host strain is normally sensitive to the antibiotic). The method is also useful where specific mutant cells are available, as the technique of **complementation** can be employed, where the cloned DNA provides the function that is absent from the mutant.

There are three requirements for this approach to be successful. Firstly, a mutant strain that is deficient in the particular gene that is being sought must be available. Secondly, a suitable selection medium is required, on which the specific recombinants will grow. The final requirement, which is often the limiting step as far as this method is concerned, is that the gene sequence must be expressed in the host cell to give a functional product that will complement the mutation. This is not a problem if, for example, *E. coli* is used to select cloned *E. coli* genes, as the cloned sequences will obviously function in the host cells. This approach has been used most often to select genes that specify nutritional requirements, such as enzymes of the various biosynthetic pathways. Thus genes of the tryptophan operon can be selected by plating recombinants on mutant cells that lack specific functions in this pathway (**auxotrophic** mutants). In some cases, complementation in *E. coli* can be used to select genes from other organisms such as yeast, if the enzymes are similar in terms of their function and they are expressed in the host cell. Complementation can also be used if mutants are available for other host cells, as is the case for yeast and other fungi.

The main limitation to direct selection is that it is generally not suitable for selecting genes from higher eukaryotes. One exception is the gene for mouse dihydrofolate reductase (DHFR), which has been cloned by selection in *E. coli* using the drug trimethoprim in the selection medium. Cells containing the mouse DHFR gene were resistant to the drug, and were therefore selected on this basis.

7.1.4 Other genetic selection methods

Although the methods outlined above represent some of the ways by which genetic selection and screening can be used to detect the presence of recombinants, there are many other examples of such techniques. These are often dependent on the use of a particular vector/host combination, which enables exploitation of the genetic characteristics of the system. Two examples will be used to illustrate this approach; many others can be found in some of the texts listed in Suggestions for further reading.

The use of the cI repressor system of λgt10 can be extended to provide a powerful **selection** system if the vector is plated on a mutant strain of *E. coli* that produces lysogens at a high frequency. Such strains are designated *hfl* (***h**igh **f**requency of **l**ysogeny*), and any phage that encodes a functional cI repressor will form lysogens on these hosts. These lysogens will be immune to further infection by phage. DNA fragments are inserted into the λgt10 vector at a restriction site in the *cI* gene. This inactivates the gene, and thus only recombinants (genotypically cI^-) will form plaques.

A second example of genetic selection based on phage/host characteristics is the Spi selection system that can be used with vectors such as EMBL4. Wild-type λ will not grow on cells that already carry a phage, such as phage P2, in the lysogenic state. Thus the λ phage is said to be Spi^+ (***s**ensitive to **P**2 **i**nhibition*). The Spi^+ phenotype is dependent on the *red* and *gam* genes of λ, and these are arranged so that they are present on the stuffer fragment of EMBL4. Thus recombinants, which lack the stuffer fragment, will be red^- gam^- and will therefore be phenotypically Spi^-. Such recombinants will form plaques on P2 lysogens, whereas non-recombinant phage that are red^+ gam^+ will retain the Spi^+ phenotype and will not form plaques.

7.2 Screening using nucleic acid hybridisation

General aspects of nucleic acid hybridisation are described in section 3.4. It is a very powerful method of screening clone banks, and is one of the key techniques in gene manipulation. The production of a cDNA or genomic DNA library is often termed the 'shotgun' approach, as a large number of essentially random recombinants is generated. By using a defined nucleic acid **probe**, such libraries can be screened and the clone(s) of interest identified. The conditions for hybridisation are now well established, and the only limitation to the method is the availability of a suitable probe.

7.2.1 Nucleic acid probes

The power of nucleic acid hybridisation lies in the fact that complementary sequences will bind to each other with a very high degree of fidelity (see Fig. 3.5). In practice this depends on the degree of **homology** between the hybridising sequences, and usually the aim is to use a probe that has been derived from the same source as the target DNA. However, under certain conditions sequences that are not 100% homologous can be used to screen for a particular gene, as may be the case if a probe from one organism is used to detect clones prepared using DNA from a second organism. Such **heterologous** probes have been extremely useful in identifying many genes from different sources.

There are three main types of DNA probe, these being (i) cDNA, (ii) genomic DNA and (iii) oligonucleotides. Alternatively, RNA probes can be used if these are suitable. The availability of a particular probe will depend on what is known about the target gene sequence. If a cDNA clone has already been obtained and identified, the cDNA can be used to screen a genomic library and isolate the gene sequence itself. Alternatively, cDNA may be made from mRNA populations and used without cloning the cDNAs. This is often used in what is known as the 'plus/minus' method of screening. If the clone of interest contains a sequence that is expressed only under certain conditions, probes may be made from mRNA populations from cells that are expressing the gene (the plus probe) and from cells that are not expressing the gene (the minus probe). By carrying out duplicate hybridisations, the clones can be identified by their different patterns of hybridisation with the plus and minus probes.

Genomic DNA probes are usually fragments of cloned sequences that are used either as heterologous probes or to identify other clones that contain additional parts of the gene in question. This is an important part of the techniques known as **chromosome walking** and **chromosome jumping**, and can enable the identification of overlapping sequences which, when pieced together, enable long stretches of DNA to be characterised.

The use of oligonucleotide probes is possible where some amino acid sequence data are available for the protein encoded by the target gene. Using the genetic code, the likely gene sequence can be derived and an oligonucleotide made. The degenerate nature of the genetic code means that it is not possible to predict the sequence with complete accuracy, but this is not usually a major problem, as mixed probes can be used that cover all the possible sequences. The great advantage of oligonucleotide probes is that only a short stretch of sequence is required for the probe to be useful, and thus genes for which clones are not already available can be identified by sequencing peptide fragments and constructing probes accordingly.

When a suitable probe has been obtained, it is usually labelled with ^{32}P as described in section 3.3. This produces a radioactive fragment of high specific activity that can be used as an extremely sensitive screen for the gene of interest. Alternatively, non-radioactive labelling methods may be used if desired.

7.2.2 Screening clone banks

Colonies or plaques are not suitable for direct screening, so a replica is made on either nitrocellulose or nylon filters. This can be done either by growing cells directly on the filter on an agar plate (colonies), or by 'lifting' a replica from a plate (colonies or plaques). To do this the recombinants are grown and a filter is placed on the surface of the agar plate. Some of the cells/plaques will stick to the filter, which therefore becomes a mirror image of the pattern of recombinants on the plate (Fig. 7.3). Reference marks are made so that the filters can be orientated correctly after hybridisation. The

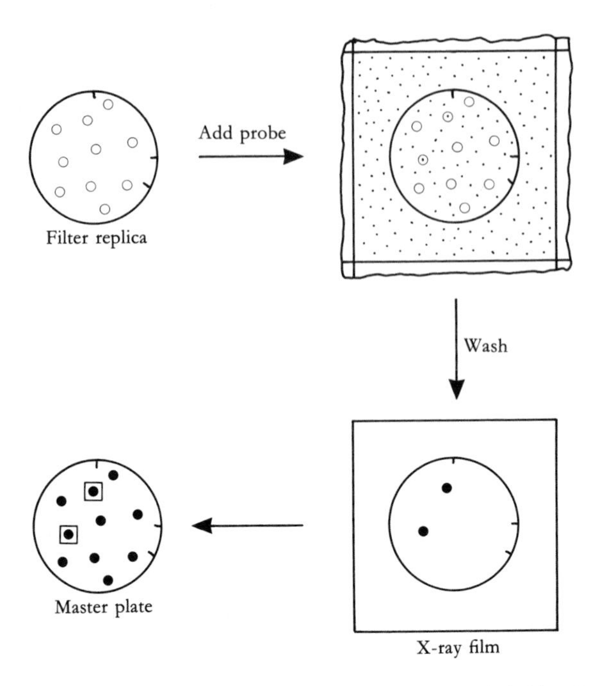

Fig. 7.3. Screening clone banks by nucleic acid hybridisation. A nitrocellulose or nylon filter replica of the master petri dish containing colonies or plaques is made. Reference marks are made on the filter and the plate to assist with correct orientation. The filter is incubated with a labelled probe, which hybridises to the target sequences. Excess or non-specifically bound probe is washed off and the filter exposed to X-ray film to produce an autoradiogram. Positive colonies (boxed) are identified and can be picked from the master plate.

filters are then processed to denature the DNA in the samples, bind this to the filter and remove most of the cell debris.

The probe is denatured (usually by heating), placed in a sealed plastic bag with the filters, and incubated at a suitable temperature to allow hybrids to form. The **stringency** of hybridisation is important, and depends on conditions such as salt concentration and temperature. For homologous probes under standard conditions incubation is usually around 65–68 °C. Time of incubation may be up to 48 h in some cases, depending on the predicted kinetics of hybridisation. After hybridisation, the filters are washed (again the stringency of washing is important) and allowed to dry. They are then exposed to X-ray film to produce an autoradiogram, which can be compared with the original plates to enable identification of the desired recombinant.

An important factor in screening genomic libraries by nucleic acid

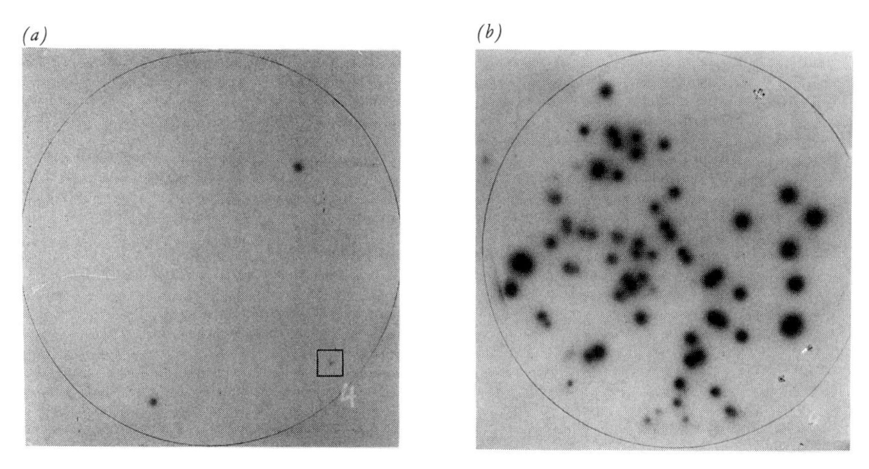

Fig. 7.4. Screening plaques at high and low densities. A radiolabelled probe was used to screen a genomic library in the λ vector EMBL3. (*a*) Initial screening was at a high density of plaques, which identified two positive plaques on this plate. The boxed area shows a false positive. (*b*) The plaques were picked from the positive areas and re-screened at a lower density to enable isolation of individual plaques. Many more positives are obtained due to the high proportion of 'target' plaques in the re-screened sample. Photograph courtesy of Dr M. Stronach.

hybridisation is the number of plaques that can be screened on each filter. Often an initial high-density screen is performed, and the plaques picked from the plate. Because of the high plaque density, it is often not possible to avoid contamination by surrounding plaques. Thus the mixture is re-screened at a much lower plaque density, which enables isolation of a single recombinant (Fig. 7.4). This approach can be important if a large number of plaques has to be screened, as it cuts down the number of filters (and hence the amount of radioactive probe) required.

7.3 Immunological screening for expressed genes

An alternative to screening with nucleic acid probes is to identify the protein product of a cloned gene by immunological methods. The technique requires that the protein is expressed in recombinants, and is most often used for screening cDNA expression libraries that have been constructed in vectors such as λgt11. Instead of a nucleic acid probe, a specific antibody is used.

Antibodies are produced by animals in response to challenge with an **antigen**, which is normally a purified protein. There are two main types of

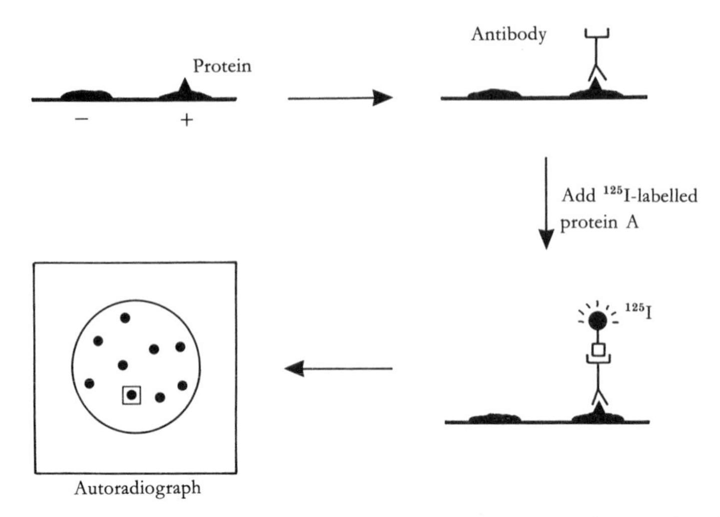

Fig. 7.5. Immunological screening for expressed genes. A filter is taken from a Petri dish containing the recombinants (usually cDNA/λ constructs). Protein and cell debris adhere to the filter. Plaques expressing the target protein (+) are indistinguishable from the others (−) at this stage. The filter is incubated with a primary antibody that is specific for the target protein. This is then complexed with radiolabelled protein A, and an autoradiogram prepared. As in nucleic acid screening, positive plaques can be identified and picked from the master plate.

antibody preparation that can be used. The most common are **polyclonal** antibodies, which are usually raised in rabbits by injecting the antigen and removing a blood sample after the immune response has occurred. The immunoglobulin fraction of the serum is purified and used as the antibody preparation (antiserum). Polyclonal antisera contain antibodies which recognise all the antigenic determinants of the antigen. A more specific antibody can be obtained by preparing **monoclonal** antibodies, which recognise a single antigenic determinant. However, this can be a disadvantage in some cases. In addition, monoclonal antibody production is a complicated technique in its own right, and good quality polyclonal antisera are usually sufficient for screening purposes.

There is a variety of methods available for immunological screening, but the technique is most often used in a similar way to 'plaque lift' screening with nucleic acid probes (Fig. 7.5). Recombinant λgt11 cDNA clones will express cloned sequences as ß-galactosidase fusion proteins, assuming that the sequence is present in the correct orientation and reading frame. The proteins can be picked up onto nitrocellulose filters and probed with the antibody. Detection can be carried out by a variety of methods, most of which use a non-specific second binding molecule such as protein A from bacteria, or a second antibody, which attaches to the specifically bound primary antibody. Detection may be by radioactive label (^{125}I-labelled protein A or second antibody) or by non-radioactive methods which produce a coloured product.

7.4 Analysis of cloned genes

Once clones have been identified by techniques such as hybridisation or immunological screening, more detailed characterisation of the DNA can begin. There are many ways of tackling this, and the choice of approach will depend on what is already known about the gene in question, and on the ultimate aims of the experiment.

7.4.1 Characterisation based on mRNA translation *in vitro*

In some cases the identity of a particular clone may require confirmation. This is particularly true when the plus/minus method of screening has been used, as the results of such a process are usually somewhat ambiguous. If the desired sequence codes for a protein, and the protein has been characterised, it is possible to identify the protein product by two methods based on translation of mRNA *in vitro*. These methods are known as **hybrid-arrest translation** (HART) and **hybrid-release translation** (HRT). A comparison of HART and HRT is shown in Fig. 7.6.

Both HART and HRT rely on hybridizing cloned DNA fragments to mRNA prepared from the cell or tissue type from which the clones have been derived. In hybrid arrest, the cloned sequence blocks the mRNA and prevents its translation when placed in a system containing all the components of the translational machinery. In hybrid release, the cloned sequence is immobilised and used to select the clone-specific mRNA from the total mRNA preparation. This is then released from the hybrid and translated *in vitro*. If a radioactive amino acid (usually [^{35}S]methionine) is incorporated into the translation mixture, the proteins synthesised from the mRNA(s) will be labelled and can be detected by autoradiography or fluorography after SDS-polyacrylamide gel electrophoresis. In hybrid arrest, one protein band should be absent, whilst in hybrid release there should be a single band. Thus hybrid release gives a cleaner result than hybrid arrest, and is the preferred method.

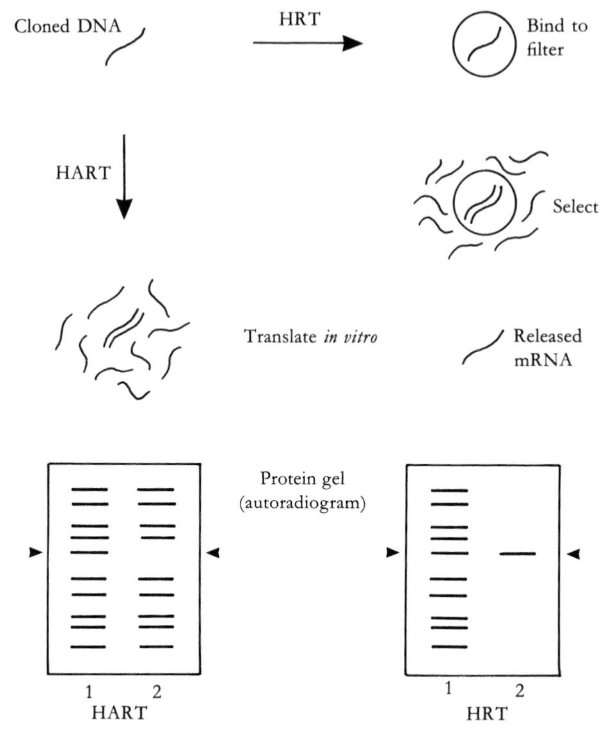

Fig. 7.6. Hybrid arrest and hybrid release translation to identify the protein product of a cloned fragment. In hybrid arrest (HART) the cloned fragment is mixed with a preparation of total mRNA. The hybrid formed effectively prevents translation of the mRNA to which the cloned DNA is complementary. After translation *in vitro*, the protein products of the translation are separated on a polyacrylamide gel. The patterns of the control (lane 1, HART gel) and test (lane 2, HART gel) translations differ by one band due to the absence of the protein encoded by the mRNA that has hybridised with the DNA. In hybrid release (HRT) the cloned DNA is bound to a filter and used to select its complementary mRNA from total mRNA. After washing to remove unbound mRNAs and releasing the specifically bound mRNA from the filter, translation *in vitro* generates a single band (lane 2, HRT gel) as opposed to the multiple bands of the control (lane 1, HRT gel). In both cases the identity of the protein (and hence the gene) can be determined by examination of the protein gels. The protein band of interest is arrowed.

7.4.2 Restriction mapping

Obtaining a restriction map for cloned fragments is usually essential before additional manipulations can be carried out. This is particularly important where phage or cosmid vectors have been used to clone large pieces of DNA. If a restriction map is available, smaller fragments can be isolated and used for various procedures including sub-cloning into other vectors, the preparation of probes for chromosome walking, and DNA sequencing.

The basic principle of restriction mapping is outlined in section 4.1.3. In practice, the cloned DNA is usually cut with a variety of restriction enzymes to determine the number of fragments produced by each enzyme. If an enzyme cuts the fragment at frequent intervals it will be difficult to decipher the restriction map, so enzymes with multiple cutting sites are best avoided. Enzymes that cut the DNA into two to four pieces are usually chosen for initial experiments. By performing a series of single and multiple digests with a range of enzymes, the complete restriction map can be pieced together. This provides the essential information required for more detailed characterisation of the cloned fragment.

7.4.3 Blotting techniques

Although a clone may have been identified and its restriction map determined, this information in itself does not provide much of an insight into the fine structure of the cloned fragment and the gene that it contains. Ultimately the aim may be to obtain the gene sequence (see section 7.4.4), but it is usually not sensible to begin sequencing straight away. If, for example, a 20 kb fragment of genomic DNA has been cloned in a λ replacement vector, and the gene of interest is only 2 kb in length, then much effort would be wasted by sequencing the entire clone. In many experiments it is therefore essential to determine which parts of the original clone contain the regions of interest. This can be done by using a variety of methods based on blotting nucleic acid molecules onto membranes, and hybridising with specific probes. Such an approach is in some ways an extension of clone identification by colony or plaque hybridisation, with the refinement that information about the structure of the clone is obtained.

The first blotting technique was developed by Ed Southern, and is eponymously known as **Southern blotting**. In this method fragments of DNA, generated by restriction digestion, are subjected to agarose gel electrophoresis. The separated fragments are then transferred to a nitrocellulose or nylon membrane by a 'blotting' technique. The original method used capillary blotting, as shown in Fig. 7.7. Although other methods such as vacuum blotting and electroblotting have been devised, the original method is still used extensively. Blots are often set up with whatever is at hand, and precarious-looking versions of the blotting apparatus are a common sight in many laboratories.

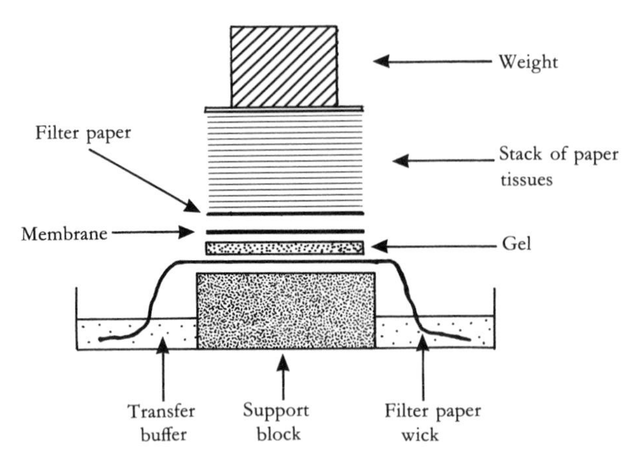

Fig. 7.7. Blotting apparatus. The gel is placed on a filter paper wick and a nitrocellulose or nylon filter placed on top. Further sheets of filter paper and paper tissues complete the set up. Transfer buffer is drawn through the gel by capillary action, and the nucleic acid fragments are transferred out of the gel and onto the membrane.

When the fragments have been transferred from the gel and bound to the filter, it becomes a replica of the gel. The filter can then be hybridised with a radioactive probe in a similar way to colony or plaque filters. As with all hybridisation, the key is the availability of a suitable probe. After hybridisation and washing, the filter is exposed to X-ray film and an autoradiogram prepared, which provides information on the structure of the clone. An example of the use of Southern blotting in clone characterisation is shown in Fig. 7.8.

Although Southern blotting is a very simple technique, it has many applications, and has been an invaluable method in gene analysis. The same technique can also be used with RNA, as opposed to DNA, and in this case is known as **Northern blotting**. It is most useful in determining hybridisation patterns in mRNA samples, and can be used to determine which regions of a cloned DNA fragment will hybridise to a particular mRNA. However, it is more often used as a method of measuring transcript levels during expression of a particular gene.

There are two further variations on the blotting theme. If nucleic acid samples are not subjected to electrophoresis, but are spotted onto the filters, hybridisation can be carried out as for Northern and Southern blots. This technique is known as **dot-blotting**, and is particularly useful in obtaining quantitative data in the study of gene expression (see section 8.1.3). The final technique is known as **Western blotting**, and this involves

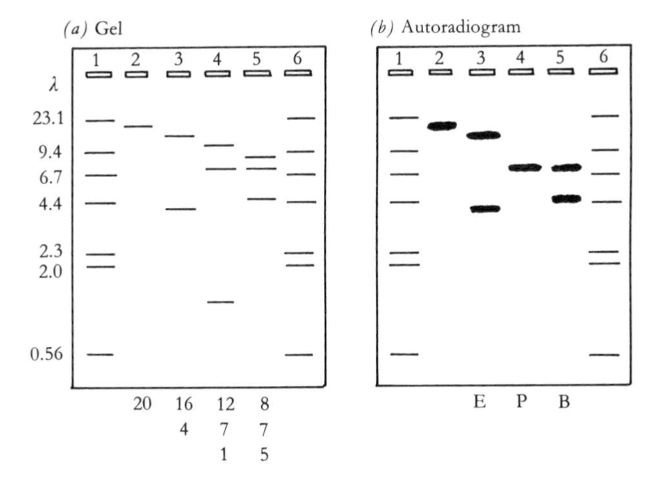

Fig. 7.8. Southern blotting. A hypothetical 20 kb fragment from a genomic clone is under investigation. A cDNA copy of the mRNA is available for use as a probe. (*a*) The gel pattern of fragments produced by digestion with various restriction enzymes; (*b*) the autoradiogram resulting from the hybridisation. Lanes 1 and 6 contain λ *Hin*dIII markers, sizes as indicated. These have been marked on the autoradiogram for reference. The intact fragment (lane 2) runs as a single band to which the probe hybridises. Lanes 3,4 and 5 were digested with *Eco*RI (E), *Pst*I (P) and *Bam*HI (B). Fragment sizes are indicated under each lane in (*a*). The results of the autoradiography show that the probe hybridizes to two bands in the *Eco*RI and *Bam*HI digests, therefore the clone must have internal sites for these enzymes. The *Pst*I digest shows hybridization to the 7 kb fragment only. This might therefore be a good candidate for subcloning, as the gene may be located entirely on this fragment.

the transfer of electrophoretically separated protein molecules to membranes. The membrane is then probed with an antibody to detect the protein of interest, in a similar way to immunological screening of plaque lifts from expression libraries.

7.4.4 DNA sequencing

The development of rapid methods for sequencing DNA, as outlined in section 3.6, has meant that this task has now become routine practice in most laboratories where cloning is carried out. Sequencing a gene provides much useful information about coding sequences, control regions and other features such as intervening sequences. Thus full characterisation of a gene will inevitably involve sequencing, and a suitable strategy must be devised to enable this to be achieved most efficiently. The complexity of a sequencing strategy depends on a number of factors, the main one being the length of the fragment that is to be sequenced. Most sequencing methods enable about 300–400 bases to be read from a sequencing gel. If the DNA is only a few hundred base-pairs long, it can probably be sequenced in a single step. However, it is more likely that the sequence will be several kilobase-pairs in length, and thus sequencing is more complex.

There are basically two ways of tackling large sequencing projects. Either a random or 'shotgun' approach is used, or an ordered strategy is devised in which the location of each fragment is known prior to sequencing. In the shotgun method, random fragments are produced and sequenced. Assembly of the complete sequence relies on there being sufficient overlap between the sequenced fragments to enable computer matching of sequences from the raw data.

An ordered sequencing strategy is usually more efficient than a random fragment approach. There are several possible ways of generating defined fragments for sequencing. Examples include: (i) isolation and sub-cloning of defined restriction fragments and (ii) generation of a series of sub-clones in which the target sequence has been progressively deleted by nucleases. If defined restriction fragments are used, the first requirement is for a detailed restriction map of the original clone. Using this, suitably sized fragments can be identified and sub-cloned into a sequencing vector such as M13 or pBluescript. Each sub-clone is then sequenced, usually by the dideoxy method (see section 3.6.2). Both strands of the DNA should be sequenced independently, so that any anomalies can be spotted and re-sequenced if necessary. The complete sequence is then assembled by computer. This is made easier if overlapping fragments have been isolated for sub-cloning, as the regions of overlap enable adjoining sequences to be identified easily.

By devising a suitable strategy and paying careful attention to detail, it is possible to derive accurate sequence data from most cloned fragments. The task of sequencing a long stretch of DNA is not trivial, but it is now such an integral part of gene manipulation technology that most gene-cloning projects involve sequence determination at some point.

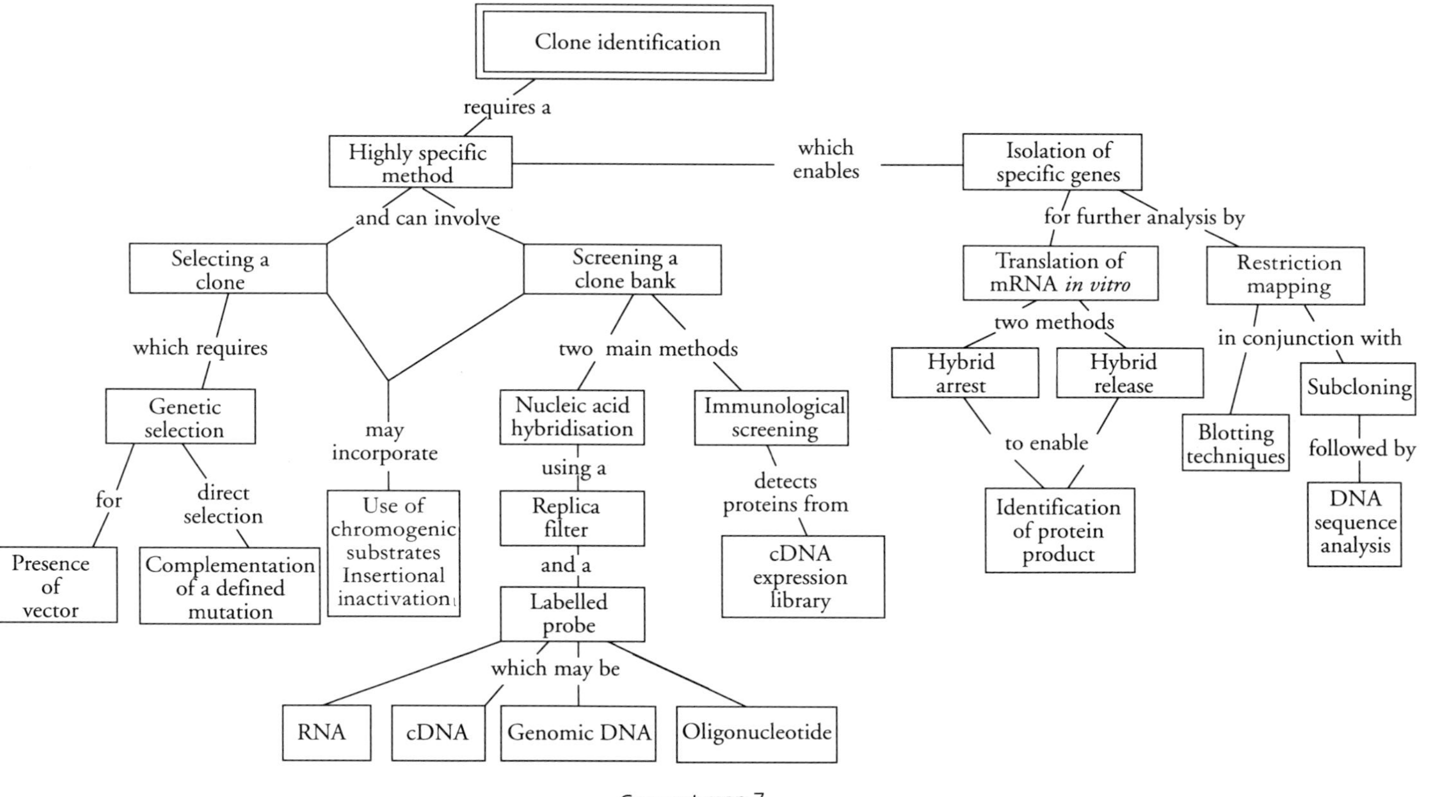

Concept map 7.

8

Genetic engineering in action

In the preceding chapters I examined the basic techniques that enable genes to be isolated and characterised. In this chapter I take a look at some of the applications of gene manipulation. Of necessity this will be a highly selective treatment, the aim being to give some idea of the immense scope of the subject whilst trying to include some detail about the various applications.

In many ways genetic engineering has undergone a shift in emphasis over the past few years, away from the technical problems that had to be solved before the technology became 'user friendly' enough for widespread use. Gene manipulation is now being used as a tool to address many diverse biological problems that were previously intractable, and the applications of the subject appear at times to be limited only by the imagination of the scientists who use it.

8.1 Analysis of gene structure and function

In terms of 'pure' science, the major impact of gene manipulation has been in the field of gene structure and expression. Although the contribution of classical genetic analysis should not be underestimated, much of the fine detail regarding gene structure and expression remained a mystery until the techniques of gene cloning enabled the isolation of individual genes.

As discussed in section 7.4, many of the techniques used to characterise cloned DNA sequences provide information about gene structure, with one of the aims of most experiments being the determination of the gene sequence. However, even when the sequence is available, there is still much work to be done to interpret the various structural features of the sequence in the context of their function *in vivo*. In this section I extend the discussion of gene analysis to include some of the methods used to investigate gene structure and function.

8.1.1 A closer look at sequences

Computer analysis of DNA sequences can provide much useful information about the structure and organisation of genes. Computers are ideally suited to this task of sequence analysis, which requires that fairly simple (but repetitive) operations are carried out quickly and accurately. Even a short sequence is tedious to analyse without the help of a computer, and there is always the possibility of error due to misreading of the sequence or to a loss in concentration.

When a gene sequence has been determined, there are a number of things that can be done with the information. Searches can be made for regions of interest, such as promoters, enhancers etc., and for sequences that code for proteins. Restriction maps can be generated easily, and printed in a variety of formats. The sequence can be compared with others from different organisms and the degree of homology between them may be determined, which can assist in studying the phylogenetic relationships between groups of organisms.

Although computer analysis of a sequence is a very useful tool, it usually needs to be backed up with experimental evidence of structure or function. For example, if a previously unknown gene is being characterised, it will be necessary to carry out experiments to determine where the important regions of the gene are. Usually such experiments confirm the function inferred from the sequence analysis, although sometimes new information is generated. Thus it is important that the computational and experimental sides of sequence analysis are used in concert.

8.1.2 Finding important regions of genes

One of the key aspects in the control of gene expression concerns protein/DNA interaction. Thus it is important to find the regions of a sequence to which the various types of regulatory proteins will bind. A relatively simple way to do this is to prepare a restriction map of the cloned DNA and generate a set of restriction fragments. The protein under investigation (perhaps RNA polymerase, a repressor protein or some other regulatory molecule) is added and allowed to bind to its site. If the fragments are then subjected to electrophoresis, the DNA/protein hybrid will run more slowly than a control fragment without protein, and can be detected by its reduced mobility. This technique is known as **gel retardation**, and it provides information about the location of particular binding sites on DNA molecules.

Although gel retardation is a useful technique, its accuracy is limited by the precision of the restriction map and the sizes of fragments that are generated. A much more precise way of identifying regions of protein binding is the technique of **DNA footprinting** (sometimes called the **DNase protection** method). The technique is elegantly simple, and relies on the fact that a region of DNA that is complexed with a protein will not be susceptible to attack by DNase I (Fig. 8.1). The DNA fragment under investigation is radiolabelled and mixed with a suspected regulatory protein. DNase I is then added so that limited digestion occurs; on average, one DNase cut per molecule is achieved. Thus a set of nested fragments will be generated, and these can be run on a sequencing gel. The region that is protected from DNase digestion gives a 'footprint' of the binding site within the molecule.

It is often necessary to locate the start site of transcription for a particular gene, and this may not be apparent from the gene sequence data. Two methods can be used to locate the T_C start site, these being **primer extension** and **S_1 mapping**. In primer extension a cDNA is synthesised from a primer that hybridises near the 5′ end of the mRNA. By sizing the fragment that is produced, the 5′ terminus of the mRNA can be identified. If a parallel sequencing reaction is run using the genomic clone and the same primer, the T_C start site can be located on the gene sequence. In S_1 mapping the genomic fragment that includes the T_C start site is labelled and used as a probe. The fragment is hybridised to the mRNA and the hybrid then digested with single-strand specific S_1 nuclease. The length of the protected fragment will indicate the location of the T_C start site relative to the end of the genomic restriction fragment.

8.1.3 Investigating gene expression

Recombinant DNA technology can be used to study gene expression in two main ways. Firstly, genes that have been isolated and characterised can be

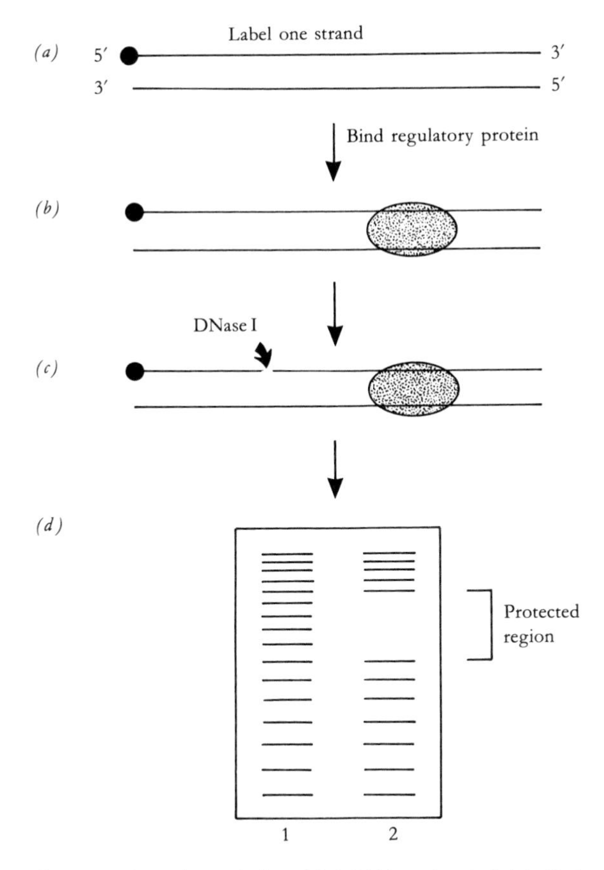

Fig. 8.1. DNA footprinting. (*a*) A DNA molecule is labelled at one end with ^{32}P. (*b*) The suspected regulatory protein is added and allowed to bind to its site. A control reaction without protein is also set up. (*c*) DNase I is used to cleave the DNA strand. Conditions are chosen so that on average only one nick will be introduced per molecule. The region protected by the bound protein will not be digested. Given the large number of molecules involved, a set of nested fragments will be produced. (*d*) The reactions are then run on a sequencing gel. When compared to the control reaction (lane 1), the test reaction (lane 2) indicates the position of the protein on the DNA by its 'footprint'.

modified and the effects of the modification studied. Secondly, probes that have been obtained from cloned sequences can be used to determine the level of mRNA for a particular protein under various conditions. These two approaches, and extensions of them, have provided much useful information about how gene expression is regulated in a wide variety of cell types.

One method of modifying genes to determine which regions are

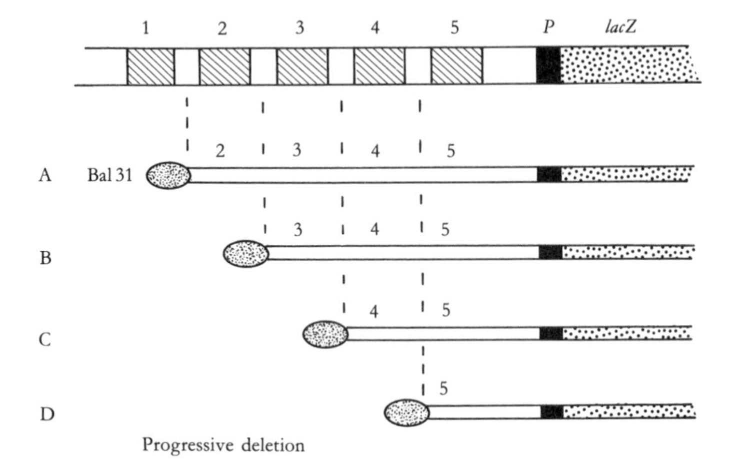

Fig. 8.2. Deletion analysis in the study of gene expression. In this hypothetical example a gene has five suspected upstream controlling regions (1 to 5, hatched). The gene promoter is labelled *P*. Often the *lacZ* gene is used as a reporter gene for detection of gene expression using the X-gal system. Deletions are created using an enzyme such as Bal 31 nuclease. In this example four deletion constructs have been made (labelled A to D). In A, region 1 has been deleted, with progressively more upstream sequence removed in each construct so that in D regions 1,2,3 and 4 have been deleted and it retains only region 5. The effects of these deletions can be monitored by the detection of β-galactosidase activity, and thus the positions of upstream controlling elements can be determined. As an alternative to using Bal 31, restriction fragments can be removed from the controlling region.

important in controlling gene expression is to delete sequences lying upstream from the T_C start site. If this is done progressively using a nuclease such as exonuclease III or Bal 31 (see section 4.2.1), a series of deletions is generated (Fig. 8.2). The effects of the various deletions can be studied by monitoring the level of expression of the gene itself, or of a 'reporter' gene such as the *lacZ* gene. In this way regions that increase or decrease transcription can be located, although the complete picture may be difficult to decipher if multiple control sequences are involved in the regulation of transcription.

Measurement of mRNA levels is an important aspect of studying gene expression, and is most often done using cDNA probes that have been cloned and characterised. The mRNA samples for probing may be from different tissue types or from cells under different physiological conditions, or may represent a time-course if induction of a particular protein is being examined. If the samples have been subjected to electrophoresis a Northern blot can be prepared, which gives information about the size of transcripts as well as their relative abundance. Alternatively, a dot-blot can be prepared and used to provide quantitative information about transcript levels by determining the amount of radioactivity in each 'dot': this reflects the amount of specific mRNA in each sample (Fig. 8.3).

Northern and dot-blotting techniques can provide a lot of useful data about transcript levels in cells under various conditions. When considered along with information about protein levels or activities, derived from Western blots or enzyme assays, a complete picture of gene expression can be built up.

8.2 Making proteins

The synthesis and purification of proteins from cloned genes is one of the most important aspects of genetic manipulation, particularly where valuable therapeutic proteins are concerned. Many such proteins have already been produced by recombinant DNA techniques, with examples including insulin, human growth hormone, interferons, ß-endorphin and factor VIII.

In protein production there are two aspects which require optimisation, these being: (i) the biology of the system and (ii) the production process itself. Careful design of both these aspects is required if the overall process is to be commercially viable, which is necessary if large-scale production and marketing of the protein is the aim.

8.2.1 Native and fusion proteins

For efficient expression of cloned DNA, the gene must be inserted into a vector that has a suitable promoter (see section 6.4.2), and which can be introduced into an appropriate host such as *E. coli*. Although this organism is not ideal for expressing eukaryotic genes, many of the problems of using *E. coli* can be overcome by constructing the recombinant so that the expression signals are recognised by the host cell. Such signals include promoters and terminators for transcription, and ribosome binding sites (Shine–Dalgarno sequences) for translation. Alternatively, a eukaryotic host such as the yeast *S. cerevisiae* may be more suitable for certain proteins.

For eukaryotic proteins, the coding sequence is usually derived from a cDNA clone of the mRNA. This is particularly important if the gene contains introns, as these will not be processed out of the primary transcript in a prokaryotic host. When the cDNA has been obtained, a suitable vector

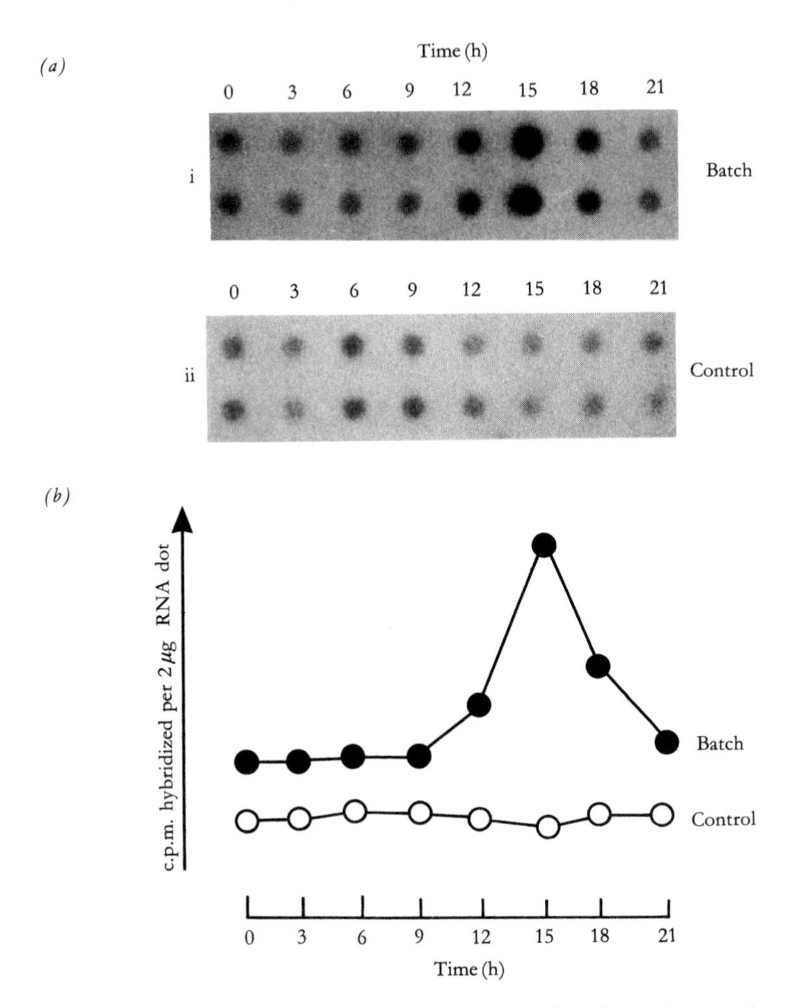

Fig. 8.3. Dot-blot analysis of mRNA levels. Samples of total RNA from synchronous cell cultures of *Chlamydomonas reinhardtii* grown under batch culture and turbidostat (control) culture conditions were spotted onto a membrane filter. The filter was probed with a radiolabelled cDNA specific for an mRNA that is expressed under conditions of flagellar regeneration. (*a*) An autoradiogram was prepared after hybridisation. Batch conditions (i) show a periodic increase in transcript levels with a peak at 15 h. Control samples (ii) show constant levels. Data shown in (*b*) were obtained by counting the amount of radioactivity in each dot. This information can be used to determine the effect of culture conditions on expression of the flagellar protein. Photograph courtesy of Dr J. Schloss. From Nicholl *et al.* (1988), *Journal of Cell Science* **89**, 397–403. Copyright (1988) The Company of Biologists Limited. Reproduced with permission.

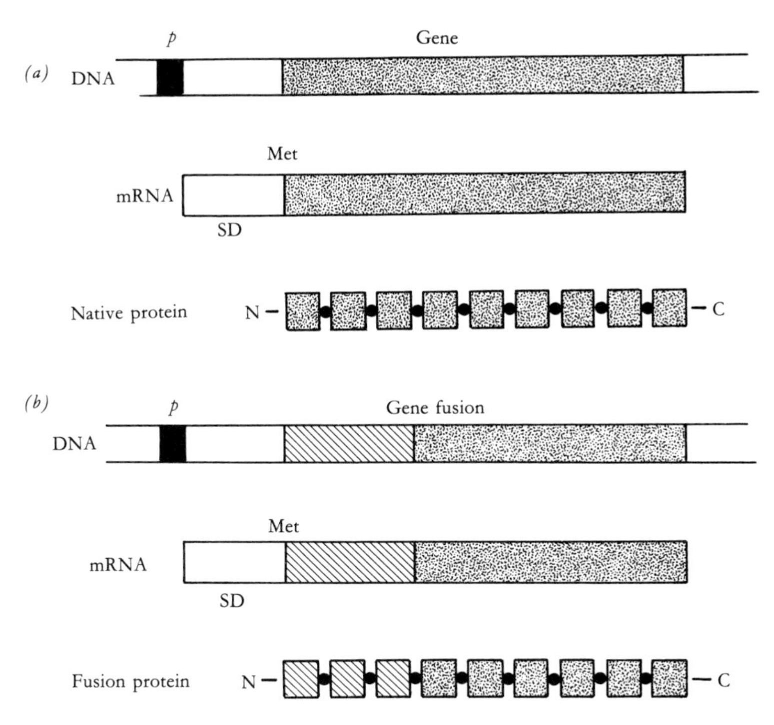

Fig. 8.4. Native and fusion proteins. (*a*) The coding sequence for the cloned gene (shaded) is not preceded by bacterial coding sequence, thus the mRNA encodes only insert-specified amino acid residues. This produces a native protein, synthesised from its own N terminus. (*b*) The gene fusion contains bacterial codons (hatched), therefore the protein contains part of the bacterial protein. In this example the first three N-terminal amino acid residues are of bacterial origin (hatched). The ribosome-binding site, or Shine–Dalgarno sequence, is marked SD.

must be chosen. Although there is a wide variety of expression vectors, there are two main categories, which produce either **native proteins** or **fusion proteins** (Fig. 8.4). Native proteins are synthesised directly from the N terminus of the cDNA, whereas fusion proteins contain short, N-terminal amino acid sequences encoded by the vector. In some cases these may be important for protein stability or secretion, and are thus not necessarily a problem. However, such sequences can be removed if the recombinant is constructed so that the fusion protein contains a methionine residue at the point of fusion. The chemical cyanogen bromide (CNBr) can be used to cleave the protein at the methionine residue, thus releasing the desired peptide. A major problem with this approach occurs if the protein contains one or more internal methionine residues, as this will result in unwanted cleavage by CNBr.

When constructing a recombinant for the synthesis of a fusion protein, it is important that the cDNA sequence is inserted into the vector in a position that maintains the correct reading frame. The addition or deletion of one or two base-pairs at the vector/insert junction may be necessary to ensure this, although there are vectors that have been constructed so that all three potential reading frames are represented for a particular vector/insert combination. Thus by using the three variants of the vector, the correct in-frame fusion can be obtained.

8.2.2 The baculovirus expression system

Baculoviruses infect insects, and do not appear to infect mammalian cells. Thus any system based on such viruses has the immediate attraction of low risk of human infection. During normal infection of insect cells, virus particles are packaged within **polyhedra**, which are nuclear inclusion bodies composed mostly of the protein **polyhedrin**. This is synthesised late in the virus infection cycle, and can represent as much as 50% of infected cell protein when fully expressed. Whilst polyhedra are required for infection of insects themselves, they are not required to maintain infection of cultured cells. Thus the polyhedrin gene is an obvious candidate for construction of an expression vector, as it encodes a late-expressed dispensable protein that is synthesised in large amounts.

The baculovirus genome is a circular double-stranded DNA molecule. Genome size is from 88–200 kb, depending on the particular virus, and the genome is therefore too large to be manipulated directly. Thus insertion of foreign DNA into the vector has to be accomplished by using an intermediate known as a **transfer vector**. These are based on *E. coli* plasmids, and carry the promoter for the polyhedrin gene (or for another viral gene) and any other essential expression signals. The cloned gene for expression is inserted into the transfer vector, and the recombinant is used to co-transfect insect cells with non-recombinant viral DNA. Homologous recombination between the viral DNA and the transfer vector results in the generation of recombinant viral genomes, which can be selected for and used to produce the protein of interest.

8.2.3 Protein engineering

One of the most exciting applications of gene manipulation lies in the field of protein engineering. This involves altering the structure of proteins *via* alterations to the gene sequence, and has become possible due to the technique of mutagenesis *in vitro*. In addition, a deeper understanding of the structural and functional characteristics of proteins has enabled workers to pinpoint the essential amino acid residues in a protein sequence. Alterations can therefore be carried out at these positions and their effects studied. The desired effect might be alteration of the catalytic activity of an enzyme by modification of the residues around the active site, an improvement in the nutritional status of a storage protein, or an improvement in the stability of a protein used in industry or medicine.

Mutagenesis *in vitro* enables specific mutations to be introduced into a gene sequence. The technique is often called **oligonucleotide-directed** or **site-directed** mutagenesis, and is elegantly simple in concept (Fig. 8.5). The requirements are a single-stranded template containing the gene to be altered, and an oligonucleotide (usually 15–30 nucleotides in length) that is complementary to the region of interest. The oligonucleotide is synthesised with the desired mutation as part of the sequence. The single-stranded template is often produced using the M13 cloning system, which produces ss DNA. The template and oligonucleotide are annealed (the mutation site will mismatch, but the flanking sequences will confer stability), and the template is then copied using DNA polymerase. This gives rise to a ds DNA which, on replication, will yield two daughter molecules, one of which will contain the desired mutation.

Identification of the mutated DNA can be carried out by hybridisation with the mutating oligonucleotide sequence, which is radiolabelled. Non-mutated DNA will retain the original mismatch, whereas the mutant will match perfectly. By washing filters of the suspected mutants at high stringency, all imperfect matches can be removed and the mutants detected by autoradiography. Even a single base-pair change can be picked up using this technique. The mutant can then be sequenced to confirm its identity.

Having altered a gene by mutagenesis, the protein is produced using an expression system. Often a vector incorporating the *lac* promoter is used, so that transcription can be controlled by the addition of IPTG. Alternatively, the λP_L promoter can be used with a temperature-sensitive λcI repressor protein, so that expression of the mutant gene is repressed at 30 °C but is permitted at 42 °C. Analysis of the mutant protein is carried out by comparison with the wild-type protein. In this way, proteins can be 'engineered' by incorporating subtle structural changes that alter their functional characteristics. This technique has great potential that is only just beginning to be exploited.

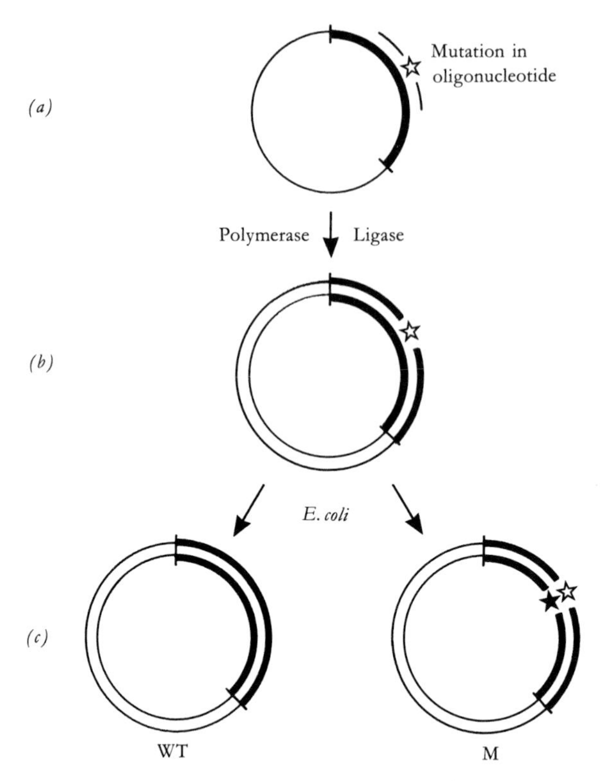

Fig. 8.5. Oligonucleotide-directed mutagenesis. (*a*) The requirement for mutagenesis *in vitro* is a single-stranded DNA template containing a cloned gene (heavy line). An oligonucleotide is synthesised that is complementary to the part of the gene that is to be mutated (but which incorporates the desired mutation). This is annealed to the template (the mutation is shown as an open star). (*b*) The molecule is made double-stranded in a reaction using DNA polymerase and ligase, which produces a hybrid wild type/mutant DNA molecule with a mismatch in the mutated region. (*c*) On introduction into *E. coli* the molceule is replicated, thus producing double-stranded copies of the wild type (WT) and mutant (M) forms. The mutant carries the original mutation and its complementary base or sequence (filled star).

8.3 Transgenic plants

The production of a **transgenic** organism involves altering the genetic makeup of that organism, and it is this area that is likely to cause the greatest public concern about genetic engineering. In addition, the scientific and technical problems associated with genetic engineering in higher organisms are often difficult to overcome, largely due to the size of the genome in such organisms, and to the fact that the development of plants and animals is an extremely complex process that is only partially understood at present. Despite these difficulties, methods for the generation of transgenic plants and animals are now well established, and the use of transgenic organisms is likely to become a major influence in the biotechnology industry.

8.3.1 Why transgenic plants?

For thousands of years humans have manipulated the genetic characteristics of plants by selective breeding. This approach has been extremely successful, and will continue to play a major part in agriculture. However, classical plant breeding programmes rely on being able to carry out genetic crosses between individual plants. Such plants must be sexually compatible (which usually means that they have to be closely related), and thus it has not been possible to combine genetic traits from widely differing species. The advent of genetic engineering has removed this constraint, and has given the agricultural scientist a very powerful way of incorporating defined genetic changes into plants. Such changes are often aimed at improving the productivity and 'efficiency' of crop plants, both of which are important to help feed and clothe the increasing world population.

There are many diverse areas of plant genetics, biochemistry, physiology and pathology involved in the genetic manipulation of plants. Some of the prime targets for the improvement of crop plants are listed in Table 8.1. In many of these, success has already been achieved to some extent. However, many people are concerned about the possible ecological effects of the release of genetically engineered organisms into the environment, and there is much debate about this aspect. The truth of the matter is that we simply do not know what the consequences might be – a very small alteration to the balance of an ecosystem, caused by a more vigorous or disease-resistant plant, might have a considerable knock-on effect over a long timescale.

As in other areas, successful genetic manipulation of plants requires: (i) methods for introducing genes into plants, and (ii) a detailed knowledge of the molecular genetics of the system that is being manipulated. In many cases the latter is the limiting factor, particularly where the characteristic under study involves many genes (a **polygenic** trait). However, despite the problems, plant genetic manipulation is already having a considerable impact on agriculture.

Table 8.1. *Possible targets for crop plant improvement*

Target	Benefit(s)
Disease, Herbicide, Insect, Virus resistance	Improve productivity of crops and reduce their loss due to biological agents
Cold, Drought, Salt tolerance	Permit growth of crops in areas that are physically unsuitable at present
Reduction of photorespiration	Increase efficiency of energy conversion
Nitrogen fixation	Confer ability to fix atmospheric nitrogen to a wider range of species
Nutritional value	Improve nutritional value of storage proteins by protein engineering
Storage properties	Extend shelf-life of fruits and vegetables
Consumer appeal	Make fruits and vegetables more appealing with respect to colour, shape, size etc.

8.3.2 Ti plasmids as vectors for plant cells

Introducing cloned DNA into plant cells is now routine practice in many laboratories worldwide. A number of methods can be used to achieve this, including physical methods such as microinjection or biolistic DNA delivery. Alternatively a biological method can be used in which the cloned DNA is incorporated into the plant by a vector. Although plant viruses such as **calumoviruses** or **geminiviruses** may be useful as vectors, there are several problems with these systems. Currently the most widely used plant cell vectors are based on the Ti plasmid of *Agrobacterium tumefaciens*, which is a soil bacterium that is responsible for crown gall disease. The bacterium infects the plant through a wound in the stem, and a tumour of cancerous tissue develops at the crown of the plant.

The agent responsible for the formation of the crown gall tumour is not the bacterium itself, but a plasmid known as the Ti (**T**umour **i**nducing) plasmid. Ti plasmids are large, ranging from 140 kb to 235 kb. In addition to the genes responsible for tumour formation, the Ti plasmids carry genes for virulence functions and for the synthesis and utilisation of unusual amino acid derivatives known as **opines**. Two main types of opine are

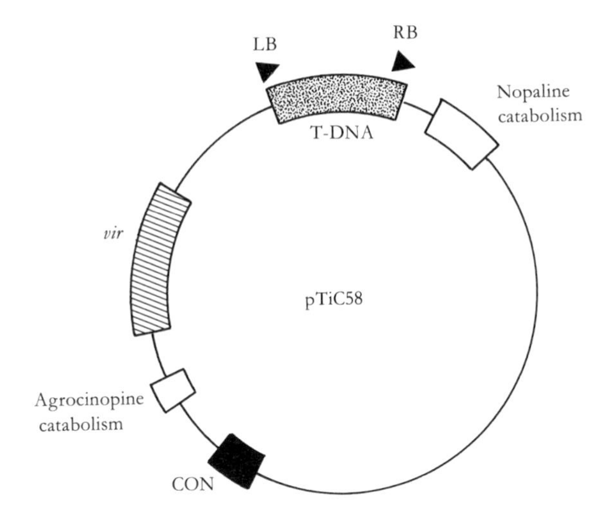

Fig. 8.6. Map of the nopaline plasmid pTiC58. Regions indicated are the T-DNA (shaded), which is bordered by left and right repeat sequences (LB and RB); the genes for nopaline and agrocinopine catabolism, and the genes specifying virulence (*vir*). The CON region is responsible for conjugative transfer. From Old and Primrose (1989), *Principles of Gene Manipulation*, 4th edition, Blackwell. Reproduced with permission.

commonly found, these being **octopine** and **nopaline**, and Ti plasmids can be characterised on this basis. A map of a nopaline Ti plasmid is shown in Fig. 8.6.

The region of the Ti plasmids responsible for tumour formation is known as the **T-DNA**. This is some 15–30 kb in size, and also carries the gene for octopine or nopaline synthesis. On infection, the T-DNA becomes integrated into the plant cell genome, and is therefore a possible avenue for the introduction of foreign DNA into the plant genome. Integration can occur at many different sites in the plant genome, the choice being apparently random. Nopaline T-DNA is a single segment, whereas octopine DNA is arranged as two regions known as the left and right segments. The left segment is similar in structure to nopaline T-DNA, and the right is not necessary for tumour formation. The structure of nopaline T-DNA is shown in Fig. 8.7. Genes for tumour morphology are designated *tms* ('shooty' tumours), *tmr* ('rooty' tumours) and *tml* ('large' tumours). The gene for nopaline synthase is designated *nos* (in octopine T-DNA this is *ocs*, encoding octopine synthase). The *nos* and *ocs* genes are eukaryotic in character, and their promoters have been used widely in the construction of vectors that express cloned sequences.

Ti plasmids are too large to be used directly as vectors, so smaller vectors

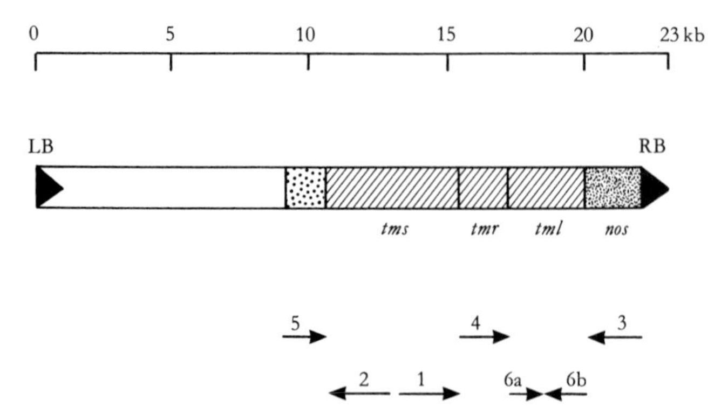

Fig. 8.7. Map of the nopaline T-DNA region. The left and right borders are indicated by LB and RB. Genes for nopaline synthase (*nos*) and tumour morphology (*tms, tmr* and *tml*) are shown. The transcript map is shown below the T-DNA map. Transcripts 1 and 2 (*tms*) are involved in auxin production, transcript 4 (*tmr*) in cytokinin production. These specify either shooty or rooty tumours. Transcript 3 encodes nopaline synthase, and transcripts 5 and 6 encode products that appear to suppress differentiation. From Old and Primrose (1989), *Principles of Gene Manipulation*, Blackwell. Reproduced with permission.

have been constructed that are suitable for manipulation *in vitro*. These vectors do not contain all the genes required for Ti-mediated gene transfer, and thus have to be used in conjunction with other plasmids to enable the cloned DNA to become integrated into the plant cell genome. Often a **tripartite** or **triparental** cross is required, where the recombinant is present in one *E. coli* strain, and a conjugation-proficient plasmid in another. A Ti plasmid derivative is present in *A. tumefaciens*. When the three strains are mixed, the conjugation-proficient 'helper' plasmid transfers to the strain carrying the recombinant plasmid, which is then mobilised and transferred to the *Agrobacterium*. Recombination then permits integration of the cloned DNA into the Ti plasmid, which can transfer this DNA to the plant genome on infection.

8.3.3 Making transgenic plants

There are two approaches to using Ti-based plasmids: (i) cointegration and (ii) the binary vector system. In the cointegration method, a plasmid based on pBR322 is used to clone the gene of interest (Fig. 8.8). This plasmid is then integrated into a Ti-based vector such as pGV3850. This carries the *vir* region (which specifies virulence), and has the left and right borders of T-DNA, which are important for integration of the T-DNA region. How-

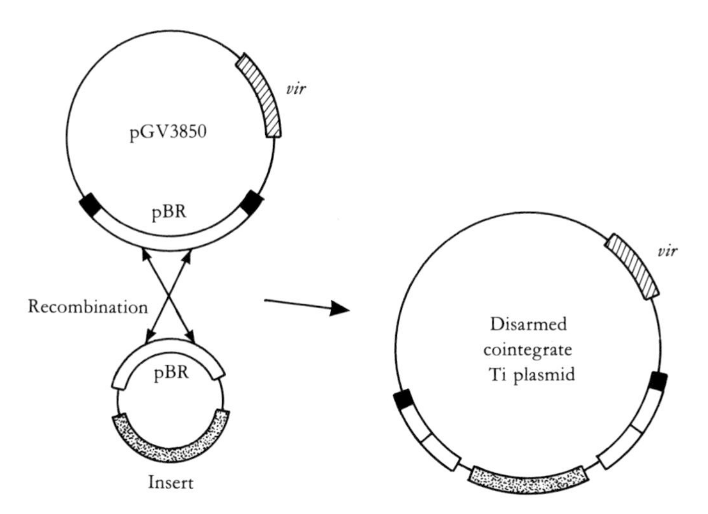

Fig. 8.8. Formation of a cointegrate Ti plasmid. Plasmid pGV3850 carries the *vir* genes, but has had some of the T-DNA region replaced with pBR322 sequences (pBR). The left and right borders of the T-DNA are present (filled regions). An insert (shaded) cloned into a pBR322-based plasmid can be inserted into pGV3850 by homologous recombination between the pBR regions, producing a disarmed cointegrate vector.

ever, most of the T-DNA has been replaced by a pBR322 sequence, which permits incorporation of the recombinant plasmid by homologous recombination. This generates a large plasmid that can facilitate integration of the cloned DNA sequence. Removal of the T-DNA has another important consequence, as cells infected with such constructs do not produce tumours and are subsequently much easier to regenerate into plants by tissue culture techniques. Ti-based plasmids lacking tumourigenic functions are known as **disarmed** vectors.

The binary vector system uses separate plasmids to supply the disarmed T-DNA (**mini-Ti** plasmids) and the virulence functions. The mini-Ti plasmid is transferred to a strain of *A. tumefaciens* (which contains a compatible plasmid with the *vir* genes) by a tri-parental cross. Genes cloned into mini-Ti plasmids are incorporated into the plant cell genome by *trans* complementation, where the *vir* functions are supplied by the second plasmid (Fig. 8.9).

When a suitable strain of *A. tumefaciens* has been generated, containing a disarmed recombinant Ti-derived plasmid, infection of plant tissue can be carried out. This is most often done using leaf discs, from which plants can be regenerated easily, and many genes have been transferred into plants by this method. The one disadvantage of the Ti system is that it does not

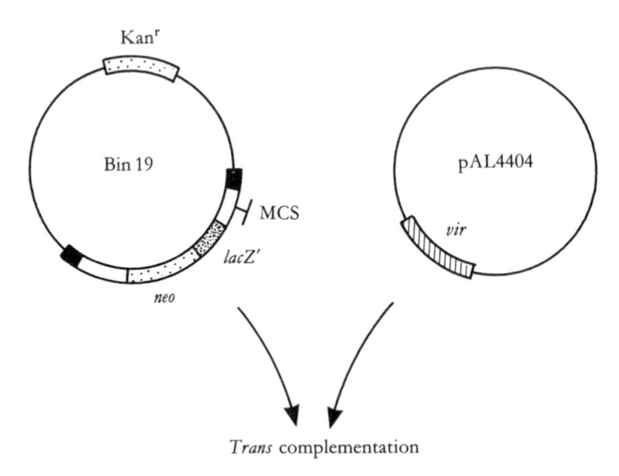

Fig. 8.9. Binary vector system based on Bin 19. The Bin 19 plasmid carries the gene sequence for the α-peptide of ß-galactosidase (*lacZ'*), downstream from a polylinker (MCS) into which DNA can be cloned. The polylinker/*lacZ'*/*neo* region is flanked by the T-DNA border sequences (filled regions). In addition genes for neomycin phosphotransferase (*neo*) and kanamycin resistance (Kan^r) can be used as selectable markers. The plasmid is used in conjunction with pAL4404, which carries the *vir* genes but has no T-DNA. The two plasmids complement each other in *trans* to enable transfer of the cloned DNA into the plant genome. After Old and Primrose (1989), *Principles of Gene Manipulation*, Blackwell. From Hoekma *et al.* (1983), *Nature (London)* **303**, 179–183, copyright (1983) Macmillan Magazines Limited; and Bevan (1984), *Nucleic Acids Research* **12**, 8711–8721, copyright (1984) IRL Press. Reproduced with permission.

normally infect monocotyledonous plants such as cereals and grasses. However, some success has been achieved when using Ti plasmids with these plants, and further work will probably extend the range of species that are amenable to Ti-mediated gene transfer.

8.4 Transgenic animals

The generation of transgenic animals is one of the most complex aspects of genetic engineering, in terms both of technical difficulty and of the ethical problems that arise. Many people who accept that the genetic manipulation of bacterial, fungal and plant species is beneficial, find difficulty in extending this acceptance when animals (particularly mammals) are involved. The need for sympathetic and objective discussion of this topic by the scientific community, the media and the general public is likely to present one of the great challenges in scientific ethics over the next few years.

8.4.1 Why transgenic animals?

Genetic engineering has already had an enormous impact on the study of gene structure and expression in animal cells, and this is one area that will continue to develop. Cancer research is one obvious example, and current investigation into the molecular genetics of the disease requires extensive use of gene manipulation technology. In the field of protein production, the synthesis of many mammalian-derived recombinant proteins is often best carried out using cultured mammalian cells, as these are sometimes the only hosts which will ensure the correct expression of such genes.

Cell-based applications such as those outlined above are an important part of genetic engineering in animals. However, the term **transgenic** is usually reserved for whole organisms, and the generation of a transgenic animal is much more complex than working with cultured cells. Many of the problems have been overcome using a variety of animals, including amphibians, fish, mice, pigs and sheep.

Transgenics can be used for a variety of purposes, covering both basic research and biotechnological applications. The study of embryological development has been extended by the ability to introduce genes into eggs or early embryos, and there is scope for the manipulation of farm animals by the incorporation of desirable traits *via* transgenesis. The use of whole organisms for the production of recombinant protein is a further possibility, and this has already been achieved in some species.

When considered on a global scale, the potential for exploitation of transgenic animals would appear to be almost unlimited. Achieving that potential is likely to be a long and difficult process in many cases, but the rewards are such that a considerable amount of money and effort has already been invested in this area.

8.4.2 Producing transgenic animals

In generating a transgenic animal, it is desirable that all the cells in the organism receive the transgene. Presence of the transgene in the germ cells of the organism will enable the gene to be passed on to succeeding generations, and this is essential if the organism is to be useful in the long term. Thus introduction of genes has to be carried out at a very early stage of development, ideally at the single-cell zygote stage. If this cannot be achieved, there is the possibility that a **mosaic** embryo will develop, in which only some of the cells carry the transgene.

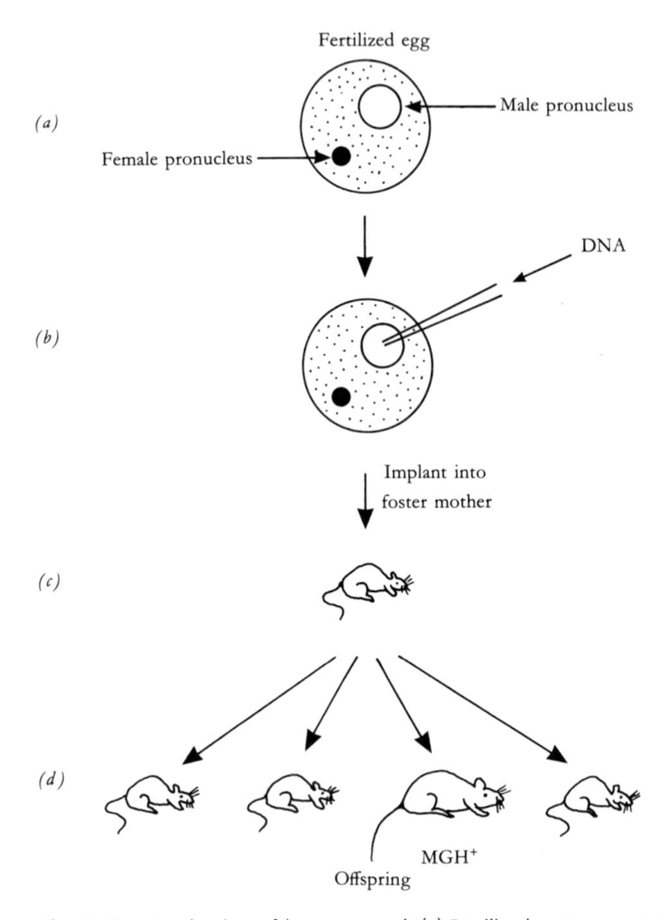

Fig. 8.10. Production of 'supermouse'. (*a*) Fertilised eggs were removed from a female and (*b*) the DNA carrying the rat growth hormone gene/mouse metallothionein promoter construct (MGH) was injected into the male pronucleus. (*c*) The eggs were then implanted into a foster mother. (*d*) Some of the pups expressed the MGH construct (MGH^+), and were larger than the normal pups.

There are several possible routes for the introduction of genes into embryos. These include: (i) direct transfection or retroviral infection of embryonic stem cells followed by introduction of these cells into an embryo at the blastocyst stage of development, (ii) retroviral infection of early embryos, and (iii) direct microinjection of DNA into zygotes or early embryo cells. Currently most success has been achieved by injecting DNA into one of the **pronuclei** of a fertilised egg, just prior to the fusion of the pronuclei (which produces the diploid zygote). This approach led to the production of the celebrated 'supermouse' in the early 1980s, which is one of the milestones of genetic engineering (Fig. 8.10).

The experiments which led to the 'supermouse' involved placing a copy of the rat growth hormone (GH) gene under the control of the mouse metallothionein (mMT) gene promoter. To create the 'supermouse', a linear fragment of the recombinant plasmid carrying the fused gene sequences (MGH) was injected into the male pronuclei of fertilised eggs (linear fragments appear to integrate into the genome more readily than circular sequences). The resulting fertilised eggs were implanted into the uteri of foster mothers, and some of the mice resulting from this expressed the growth hormone gene. Such mice grew some 2–3 times faster than control mice, and were up to twice the size of the controls.

8.4.3 Applications of transgenic animals

Introduction of growth hormone genes into other species has been carried out, notably in pigs, but in many cases there are undesirable side effects. Pigs with the bovine growth hormone gene show greater feed efficiency and have lower levels of subcutaneous fat than normal pigs. However, problems such as enlarged heart, high incidence of stomach ulcers, dermatitis, kidney disease and arthritis have demonstrated that the production of healthy transgenic farm animals is a difficult undertaking. Clearly much more work is required before genetic engineering has a major impact on animal husbandry.

The study of development is one area of transgenic research that is currently yielding much useful information. By implanting genes into embryos, features of development such as tissue-specific gene expression can be investigated. The cloning of genes from the fruit fly *Drosophila melanogaster*, coupled with the isolation and characterisation of transposable elements (P elements) that can be used as vectors, has enabled the production of stable transgenic *Drosophila* lines. Thus the fruit fly, which has been a major contributor to the field of classical genetic analysis, is now being studied at the molecular level by employing the full range of gene manipulation techniques.

In mammals, the mouse is proving to be one of the most useful model systems for investigating embryological development, and the expression of many transgenes has been studied in this organism. One such application is shown in Fig. 8.11, which demonstrates the use of the *lacZ* gene as a means of detecting tissue-specific gene expression. In this example the *lacZ* gene was placed under the control of the weak thymidine kinase (TK) promoter from herpes simplex virus (HSV), generating an HSV-TK–*lacZ* construct. This was used to probe for active chromosomal domains in the

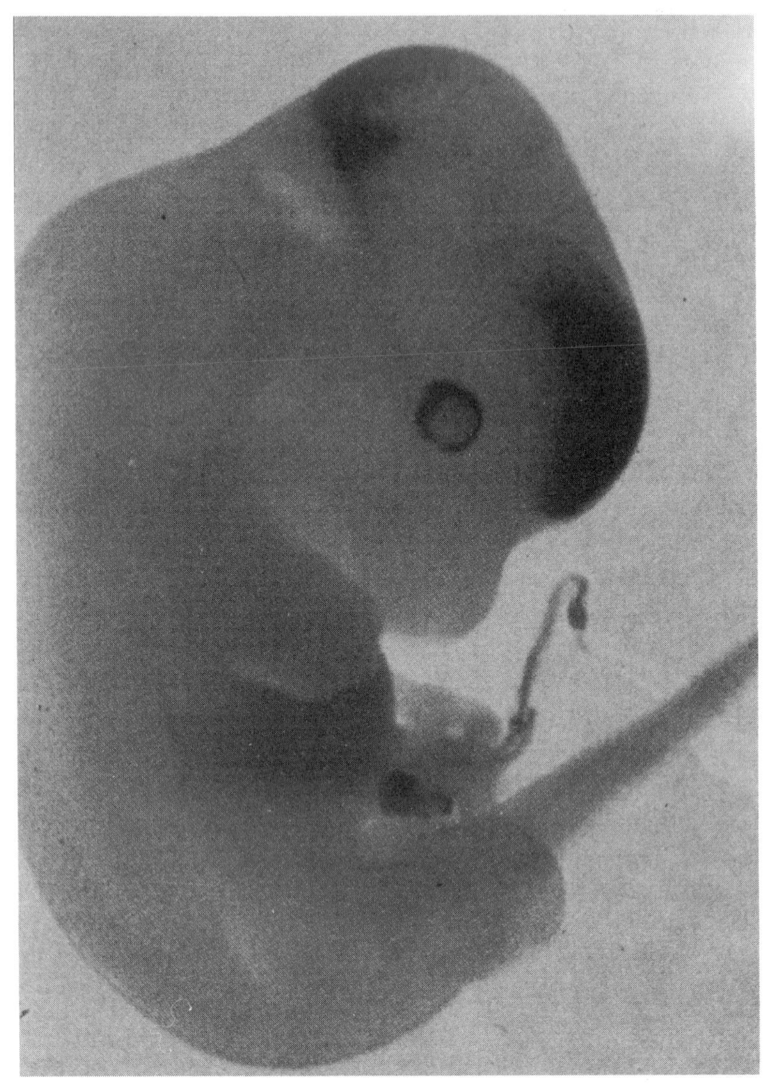

Fig. 8.11. Expression of a transgene in the mouse embryo. The ß-galactosidase (*lacZ*) coding region was placed under the control of a thymidine kinase promoter from herpes simplex virus to produce the HSV-TK–*lacZ* gene construct. This was injected into male pronuclei and transgenic mice produced. The example shows a 13-day foetus from a transgenic strain that expresses the transgene in brain tissue during gestation. Detection is by the blue colouration produced by the action of ß-galactosidase on X-gal. Thus the dark areas in the fore and hind-brain are regions where the *lacZ* gene has been expressed. Photograph courtesy of Dr S. Hettle. From Allen *et al.* (1988), *Nature (London)* **333**, 852–855. Copyright (1988) Macmillan Magazines Limited. Reproduced with permission.

developing embryos, with one of the transgenic lines showing the brain specific expression seen in Fig. 8.11.

Although the use of transgenic organisms is providing many insights into developmental processes, inserted genes may not always be expressed in exactly the same way as would be the case in normal embryos. Thus a good deal of caution is often required when interpreting results. Despite this potential problem, transgenesis is proving to be a powerful tool for the developmental biologist.

Examples of protein production in transgenics include expression of human tissue plasminogen activator (tPA) in transgenic mice, and of human blood coagulation factor IX (FIX) in transgenic sheep. In both cases the transgene protein product was secreted into the milk of lactating organisms by virtue of being placed under the control of a milk-protein gene promoter. In the mouse example the construct consisted of the regulatory sequences of the whey acid protein (WAP) gene, giving a WAP–tPA construct. Control sequences from the ß-lactoglobulin (BLG) gene were used to generate a BLG–FIX construct for expression in the transgenic sheep.

The success of these experiments opens up the exciting possibility of using transgenic animals for the production of high-value proteins. This offers an alternative to the fermentation of bacteria or yeast that contain the target gene. Given the problems in achieving correct expression and processing of some mammalian proteins in non-mammalian hosts, this may prove to be a vital development of the technology.

8.5 Spin-off technologies

In addition to mainstream branches in both pure and applied research and in the biotechnology industry, genetic manipulation has had important consequences in other areas. Some of what might be called the 'spin-off' applications of gene manipulation will be described briefly to illustrate the widespread impact of the technology.

8.5.1 Recombinant DNA technology and medicine

The diagnosis and treatment of human disease is one area in which genetic manipulation is beginning to have a considerable effect. As mentioned in section 8.2, the synthesis of therapeutic proteins by recombinant DNA methods is already well established, and the number of such proteins available is increasing steadily. In addition to proteins such as insulin, tPA, factor VIII, interferons and growth hormone, there is great interest in the development of vaccines. By utilising recombinant techniques, it is possible to develop vaccines based on selected antigenic determinants of viruses. This removes the problems that may be associated with using attenuated strains of the intact virus. In addition, it is possible to construct recombinant vaccinia viruses that express a number of foreign genes, which may enable multiple vaccinations to be achieved by a single inoculation.

The diagnosis of genetically based diseases has been revolutionised by the techniques of gene manipulation. Currently some 4000 diseases of genetic origin have been identified, and many have already been studied extensively using the retrospective technique of **pedigree analysis**. However, we now have the means to detect genetic defects by analysis of DNA samples. One important development is the use of **restriction fragment length polymorphisms** (RFLPs) as genetic markers. RFLPs are differences in the lengths of specific restriction fragments generated when DNA is digested with a particular enzyme (Fig. 8.12). They are produced when there is a variation in DNA that alters either the recognition sequence or the location of a restriction enzyme recognition site. Thus a point mutation might abolish a particular restriction site (or create a new one), whereas an insertion or deletion would alter the relative positions of restriction sites. If the RFLP lies within (or close to) the locus of a gene that causes a particular disease, it is often possible to trace the defective gene by looking for the RFLP, using the Southern blotting technique in conjunction with a probe that hybridizes to the region of interest. This approach is extremely powerful, and has already enabled many genes to be mapped to their chromosomal locations. Examples include the genes for Huntington's disease (chromosome 4), cystic fibrosis (chromosome 7), sickle-cell anaemia (chromosome 11), retinoblastoma (chromosome 13) and Alzheimer's disease (chromosome 21).

In addition to RFLP analysis, oligonucleotide probes can be synthesised and used to detect mutations in specific genes, if the gene defect is known. By using the correct probe under hybridisation conditions of high stringency, defects such as insertions, deletions and even point mutations can be detected. This approach has tremendous potential, and is constrained only by the need to know the precise nature of the gene defect.

Once genetic defects have been identified and characterised, the possibility of treating the patient arises. If the defective gene can be replaced with a functional copy that is expressed correctly, the disease caused by the defect

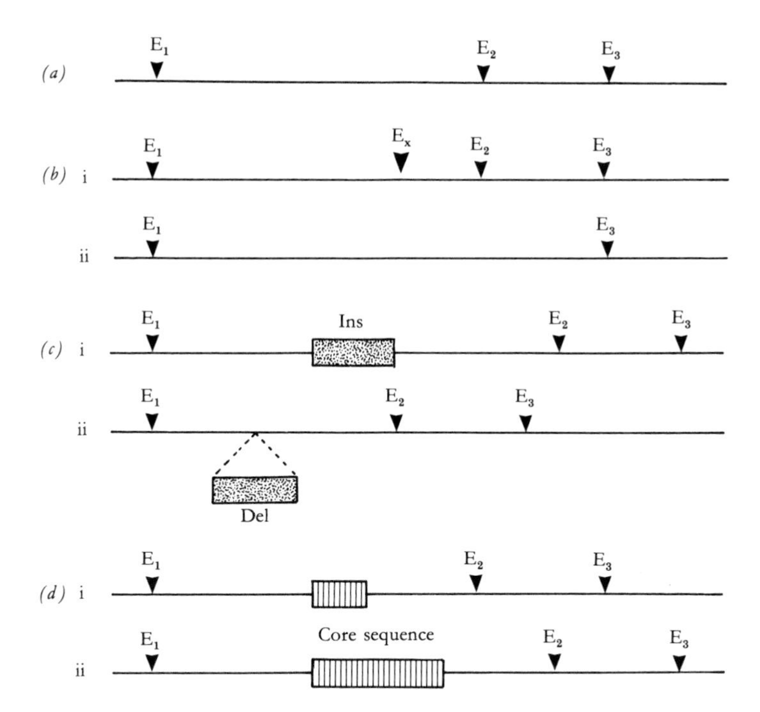

Fig. 8.12. Generation of restriction fragment length polymorphisms (RFLPs). (*a*) Consider a DNA fragment with three *Eco*RI sites (E_1,E_2 and E_3). On digestion with *Eco*RI, one of the fragments produced is $E_1 \rightarrow E_2$. RFLPs can be generated if the relative positions of these two sites are altered in any way. (*b*) The effect of point mutations. If a point mutation creates a new *Eco*RI site ((*b*)i, marked E_x), fragment $E_1 \rightarrow E_2$ is replaced by two shorter fragments, $E_1 \rightarrow E_x$ and $E_x \rightarrow E_2$. If a point mutation removes an *Eco*RI site ((*b*)ii, site E_2 removed), the fragment becomes $E_1 \rightarrow E_3$, which is longer. (*c*) The effect of insertions or deletions. If additional DNA is inserted between E_1 and E_2, (Ins in (*c*)i) fragment $E_1 \rightarrow E_2$ becomes larger. Insertions might also carry additional *Eco*RI sites, which would affect fragment lengths. If DNA is deleted (Del in (*c*)ii), the fragment is shortened. (*d*) The effect of variable numbers of repetitive core sequence motifs. This variation can be considered as a type of RFLP, with applications in genetic fingerprinting. (*d*)i has 10 copies of the core sequence, thus fragment $E_1 \rightarrow E_2$ is smaller than that shown in (*d*)ii, which has 24 copies of the core sequence. Differences in the lengths of the restriction fragments shown may be detected by Southern blotting using a suitable probe.

can be prevented. This approach is known as **gene replacement therapy** or **gene therapy**, and is one of the most promising aspects of the use of gene technology in medicine. There are two possible approaches to gene therapy: (i) introduction of the 'good' gene into the somatic cells of the affected tissue, or (ii) introduction into the reproductive (germ line) cells. These two approaches have markedly different ethical implications. Most scientists consider somatic cell gene therapy an acceptable practice, no more morally troublesome than taking an aspirin. However, tinkering with the reproductive cells, with the probability of germ line transmission, is akin to altering the gene pool of the human species, which is regarded as unacceptable by most people. Thus genetic engineering of germ cells will almost certainly be prohibited in the foreseeable future.

8.5.2 The human genome project

The biggest project currently underway in biology is the proposal to map and sequence the human genome. This is a task of almost unimaginable complexity and scale. Molecular biology has traditionally involved small groups of workers in individual laboratories, and most of the key discoveries have been made in this way. Sequencing the 3×10^9 base-pairs of the human genome is in another league altogether, and will require international collaboration on a massive scale. The idea that this could realistically be accomplished gained credibility in the mid 1980s, and by the end of the decade the project had acquired sufficient momentum to ensure that it would be supported. The initial impetus had come from the USA, and the Human Genome Project (HGP) was officially launched in October 1990. Formation of the Human Genome Organisation (HUGO) in the same year marked the birth of the project on an international scale, the role of HUGO being to co-ordinate the efforts of the many countries involved.

The genome project has many diverse component parts. There are technological considerations such as the development of automated sequencing equipment (without which the task would be all but impossible), and the need for computer systems that can cope with the vast amount of information that will be generated. The biological problems are, as might be expected, enormous. It is estimated that there are about 100 000 genes in the human genome, but this represents only about 5% of the DNA. Many scientists feel that sequencing all the DNA is a waste of time, and that effort should be concentrated on locating and sequencing the genes themselves. Others are convinced that the 'junk' DNA will yield information that will be essential in gaining a full understanding of how the genome works.

The projected timescale for the genome project will depend largely on how the technology develops, but a number of targets have been set. In addition to the human genome, several 'model genome' projects are underway, including mapping and sequencing the genomes of *E. coli*,

Saccharomyces cerevisiae, the nematode *Caenorhabditas elegans*, *Arabidopsis thaliana* (a simple plant), *Drosophila melanogaster*, mouse and pig. The period 1991–95 should see major improvements in the technology of sequencing and information handling, and the completion of low-resolution genetic and physical maps of the human genome. The next phase (1995–2000) may see some of the smaller projects completed, and more refined mapping of the human genome coupled with the generation of a considerable amount of the sequence. The complete human sequence is scheduled for completion by 2005. Although there is still much debate regarding the wisdom of undertaking the genome project, there is little doubt that it will provide an invaluable resource for biological research.

8.5.3 Questions and answers for the legal profession

There are several aspects of gene manipulation that have generated legal questions. The filing of patents is one such area. Biotechnology companies understandably wish to protect developments in which they have invested large amounts of money, but many people find difficulty in accepting that biological systems can be patented. This area has already raised some difficult questions for scientists and lawyers, and is likely to generate further controversy as more cases are dealt with.

The Human Genome Project is another area that raises a number of ethical and legal questions. Whilst the information generated will be of great benefit in many areas of science and medicine, there are fears that it might be used for what might be termed 'genetic discrimination'. If, for example, it becomes possible to determine that a certain individual is likely to develop heart disease in middle age, will insurance companies use this genome-based information when determining the premiums for life assurance? Is this any different to asking questions about family history with regard to coronary disease? This is just one hypothetical example of the possible use (misuse?) of information that may arise from the Genome Project. Many others can be envisaged, and such problems require careful consideration.

Whilst the questions raised by patent filing and the Human Genome Project might be considered as slightly unsavoury or negative consequences of gene manipulation, the development of **genetic fingerprinting** has had a major positive impact in areas of basic research, forensic science and paternity testing. The technique is based on the fact that the human genome contains polymorphic loci known as **hypervariable regions**

(HVRs). These are made up of variable number of tandem repeats of a short core sequence (see Fig. 8.12), and can be detected by using a hybridisation probe that binds to these regions. If genomic DNA is digested with a restriction enzyme that cuts outside the repeated regions, the pattern of bands obtained on hybridisation of the probe to a Southern blot gives a genetic 'fingerprint' that is unique to the individual from which the DNA was prepared.

Genetic fingerprinting enables identification of suspects in cases where samples of blood or semen have been obtained, and criminal convictions have already been made based on evidence generated by the technique. In paternity disputes, genetic fingerprinting can provide unequivocal evidence of family relationships, because the band pattern for a given individual is inherited (derived from the maternal and paternal chromosomes). Thus the fingerprint of the child will contain bands that match some of those in the maternal and paternal fingerprints; the probability of a series of matching bands being generated by chance is essentially negligible.

In addition to the applications outlined above, genetic fingerprinting is also a powerful research tool, and can be used with other organisms such as cats, dogs, birds and plants. Application of the technique in an ecological context enables problems that were previously studied by classical ecological methods to be investigated at the molecular level. This use of 'molecular ecology' is likely to become more widespread, and will have a major impact on the study of organisms in their natural environments.

8.6 What lies ahead?

Over the past 20 years the field of genetic manipulation has developed at a staggering pace, and has revolutionised the way that biological research is carried out. In looking forward to the coming 20 years, it is difficult to predict what will happen, given the immense potential of gene manipulation in areas of basic science, biotechnology and medicine. A major challenge for the scientific community is to inform and educate the public by frank and open discussion of the relevant issues. Assuming that such an approach results in widespread public acceptance, we will certainly feel the effects of genetic engineering in many areas of our lives. It is to be hoped that all these effects are beneficial, and that we can avoid the misuse of a technology that holds tremendous promise and excitement for the years ahead.

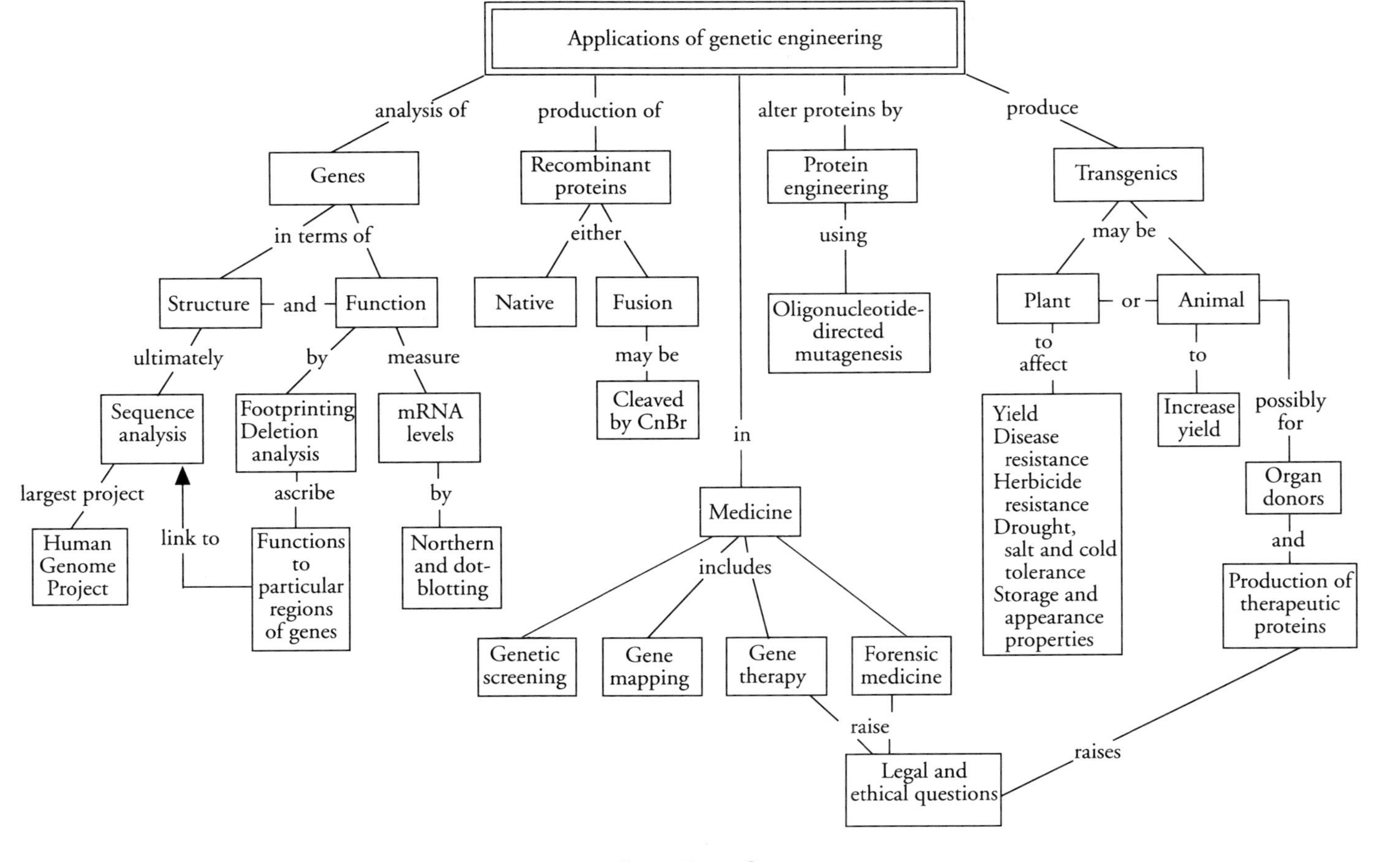

Concept map 8.

Suggestions for further reading

The following texts may be useful for further reference. All provide excellent coverage of their particular topic. In addition, the journals such as *Scientific American* and *New Scientist* often publish reviews of major topics in genetics, molecular biology and gene manipulation.

General molecular biology texts

Freifelder, D. *Molecular Biology*. 2nd edition (1987). Jones and Bartlett, Boston, USA. ISBN 0-86720-069-3, 834 pp. A comprehensive introduction to molecular biology.

Schleif, R. *Genetics and Molecular Biology* (1985). Addison-Wesley, Reading, USA. ISBN 0-201-07418-4, 626 pp. Looks at the experimental aspects of the subject, showing how key experiments helped shape the development of molecular biology.

Watson, J.D., Hopkins, N.H., Roberts, J.W., Steitz, J.A. and Weiner, A.M. *Molecular Biology of the Gene*. 4th edition (1987). Benjamin Cummings, Menlo Park, USA. ISBN 0-8053-9612-8, 1163 pp. A classic textbook. Two volumes, one dealing with general principles and one with specialised aspects of molecular biology.

Texts dealing with aspects of genetic engineering

Technical aspects

Williams, J.G. and Patient, R.K. *Genetic Engineering* (1988). IRL Press, Oxford,

UK. ISBN 1-85221-071-0, 72 pp. A short book that is a handy reference for those who already know a little about biochemistry.

Watson, J.D., Gilman, M., Witkowski, J. and Zoller, M. *Recombinant DNA*. 2nd Edition (1992). Scientific American/W.H. Freeman, New York, USA. ISBN 0-7167-2282-8, 626pp. A new edition of an early classic. A well-written and superbly illustrated introduction to genetic engineering.

Brown, T.A. *Gene Cloning – an Introduction*. 2nd Edition (1990). Chapman and Hall, London, UK. ISBN 0-412-34210-3, 286pp. An excellent introductory-level text. Easy to read, it is packed with clear illustrations.

Old, R.W. and Primrose, S.B. *Principles of Gene Manipulation – an Introduction to Genetic Engineering*. 4th Edition (1989). Blackwell, Oxford, UK. ISBN 0-632-02608-1, 438 pp. Now in its fourth edition, this is another classic textbook at advanced undergraduate level.

Winnacker, E.L. *From Genes to Clones – Introduction to Gene Technology* (1987). VCH, Weinhein, Germany. ISBN 3-527-26199-0, 634 pp. Contains a wealth of detailed information, and is perhaps the most useful advanced textbook in the field.

Kingsman, S.M. and Kingsman, A.J. *Genetic Engineering – an Introduction to Gene Analysis and Exploitation in Eukaryotes* (1988). Blackwell, Oxford, UK. ISBN 0-632-01519-7, 522 pp. Deals specifically with gene analysis in eukaryotes, covering the subject at advanced level.

Biotechnology

Marx, J.L. (ed.) *A Revolution in Biotechnology* (1989). ICSU/Cambridge University Press, Cambridge, UK. ISBN 0-521-32749-0, 227 pp. A very useful series of articles dealing with many important areas of biotechnology.

Peters, P. *Biotechnology – a Guide to Genetic Engineering* (1992). W.C. Brown, Dubuque, USA. ISBN 0-697-12063-5, 253 pp. Good coverage of the impact of genetic engineering on aspects of biotechnology, and not too technical.

Primrose, S.B. *Molecular Biotechnology*. 2nd edition (1991). Blackwell, Oxford, UK. ISBN 0-632-03053-4, 196 pp. Covers the main branches of modern biotechnology, presenting a lot of useful information.

Wider issues

Chadwick, R.F. (ed.) *Ethics, Reproduction and Genetic Control* (1990). Routledge, London, UK. ISBN 0-415-05188-6, 200 pp. Deals with the ethical implications of modern advances in genetics as applied to human reproduction.

Fincham, J.R.S. and Ravetz, J.R. *Genetically Engineered Organisms – Benefits and Risks* (1991). Open University Press, Milton Keynes, UK. ISBN 0-335-09618-2, 158 pp. Published in collaboration with the Council for Science and Society. This book provides good coverage of the applications of gene technology, and assesses the risks associated with such applications.

Nossal, G.J.V. and Coppel, R.L. *Re-shaping Life – Key Issues in Genetic Engineering.* 2nd edition (1989). Cambridge University Press, Cambridge, UK. ISBN 0-512-38969-0, 179 pp. Examines some of the key ethical issues in genetic engineering. Covers a wide range of topics without becoming too technical for the non-specialist.

Glossary

Adaptor A synthetic single-stranded non self-complementary oligonucleotide used in conjunction with a linker to add cohesive ends to DNA molecules.

Adenine (A) Nitrogenous base found in DNA and RNA.

Agarose Jelly-like matrix, extracted from seaweed, used as a support in the separation of nucleic acids by gel electrophoresis.

Alkaline phosphatase An enzyme that removes 5′-phosphate groups from the ends of DNA molecules, leaving 5′-hydroxyl groups.

Allele One of two or more variants of a particular gene.

Ampicillin (Ap) A semisynthetic ß-lactam antibiotic.

Antibody An immunoglobulin that specifically recognises and binds to an antigenic determinant on an antigen.

Anticodon The three bases on a tRNA molecule that are complementary to the codon on the mRNA.

Antigen A molecule that is bound by an antibody. Also used to describe molecules that can induce an immune response, although these are more properly described as immunogens.

Antiparallel The arrangement of complementary DNA strands, which run in different directions with respect to their 5′→3′ polarity.

Autoradiogram Image produced on X-ray film in response to the emission of radioactive particles.

Auxotroph A cell that requires nutritional supplements for growth.

Bacteriophage A bacterial virus.

Baculovirus A particular type of virus that infects insect cells, producing large inclusions in the infected cells.

Bal 31 nuclease An exonuclease that degrades both strands of a DNA molecule at the same time.

Bacterial alkaline phosphatase (BAP) See *alkaline phosphatase.*

Biolistic Refers to a method of introducing DNA into cells by bombarding them with microprojectiles, which carry the DNA.

Blunt ends DNA termini without overhanging 3′ or 5′ ends. Also known as *flush ends.*

C terminus Carboxyl terminus, defined by the –COOH group of an amino acid or protein.

CAAT box A sequence located approximately 75 base-pairs upstream from eukaryotic transcription start sites. This sequence is one of those that enhance binding of RNA polymerase.

Calf intestinal phosphatase (CIP) See *alkaline phosphatase.*

Cap A chemical modification that is added to the 5′ end of a eukaryotic mRNA molecule during post-transcriptional processing of the primary transcript.

Capsid The protein coat of a virus.

cDNA DNA that is made by copying mRNA using the enzyme reverse transcriptase.

cDNA library A collection of clones prepared from the mRNA of a given cell or tissue type, representing the genetic information expressed by such cells.

Central dogma Statement regarding the unidirectional transfer of information from DNA to RNA to protein.

Chromosome A DNA molecule carrying a set of genes. There may be a single chromosome, as in bacteria, or multiple chromosomes, as in eukaryotic organisms.

***Cis*-acting element** A DNA sequence that exerts its effect only when on the same DNA molecule as the sequence it acts on. For example, the CAAT box (q.v.) is a *cis*-acting element for transcription in eukaryotes.

Cistron A sequence of bases in DNA that specifies one polypeptide.

Clone (1) A colony of identical organisms; often used to describe a cell carrying a recombinant DNA fragment. (2) Used as a verb to describe the generation of recombinants.

Clone bank See *cDNA library, genomic library.*

Codon The three bases in mRNA that specify a particular amino acid during translation.

Cohesive ends Those ends (termini) of DNA molecules that have short complementary sequences that can stick together to join two DNA molecules. Often generated by restriction enzymes.

Competent Refers to bacterial cells that are able to take up exogenous DNA.

Complementary DNA See *cDNA.*

Complementation Process by which genes on different DNA molecules interact. Usually a protein product is involved, as this is a diffusible molecule that can exert its effect away from the DNA itself. For example, a *lacZ*$^+$ gene on a plasmid can complement a mutant (*lacZ*$^-$) gene on the chromosome by enabling the synthesis of ß-galactosidase.

Concatemer A DNA molecule composed of a number of individual pieces joined together *via* cohesive ends (q.v.).

Conjugation Plasmid-mediated transfer of genetic material from a 'male' donor bacterium to a 'female' recipient.

Consensus sequence A sequence that is found in most examples of a particular genetic element, and which shows a high degree of conservation. An example is the CAAT box (q.v.).

Copy number (1) The number of plasmid molecules in a bacterial cell. (2) The number of copies of a gene in the genome of an organism.

cos **site** The region generated when the cohesive ends of lambda DNA join together.

Cosmid A hybrid vector made up of plasmid sequences and the cohesive ends (*cos* sites) of bacteriophage lambda.

Cytosine (C) Nitrogenous base found in DNA and RNA.

Deletion Change to the genetic material caused by removal of part of the sequence of bases in DNA.

Deoxynucleoside triphosphate (dNTP) Triphosphorylated ('high energy') precursor required for synthesis of DNA, where N refers to one of the four bases (A,G,T or C).

Deoxyribonuclease (DNase) An nuclease enzyme that hydrolyses (degrades) single- and double-stranded DNA.

Deoxyribonucleic acid (DNA) A condensation heteropolymer composed of nucleotides. DNA is the primary genetic material in all organisms apart from some RNA viruses. Usually double-stranded.

Deoxyribose The sugar found in DNA.

Dideoxynucleoside triphosphate (ddNTP) A modified form of dNTP used as a chain terminator in DNA sequencing.

DNA fingerprinting See *genetic fingerprinting.*

DNA footprinting Method of identifying regions of DNA to which regulatory proteins will bind.

DNA ligase Enzyme used for joining DNA molecules by the formation of a phosphodiester bond between a 5′-phosphate and a 3′-OH group.

DNA polymerase An enzyme that synthesises a copy of a DNA template.

Dot-blot Technique in which small spots, or 'dots', of nucleic acid are immobilised on a nitrocellulose or nylon membrane for hybridisation.

Electroporation Technique for introducing DNA into cells by giving a transient electric pulse.

End labelling Adding a radioactive molecule onto the end(s) of a polynucleotide.

Endonuclease An enzyme that cuts within a nucleic acid molecule, as opposed to an exonuclease (q.v.), which digests DNA from one or both ends.

Enhancer A sequence that enhances transcription from the promoter of a eukaryotic gene. May be several thousand base-pairs away from the promoter.

Enzyme A protein that catalyses a specific reaction.

Ethidium bromide A molecule that binds to DNA and fluoresces when viewed under ultraviolet light. Used as a stain for DNA.

Eukaryotic The property of having a membrane-bound nucleus.

Exon Region of a eukaryotic gene that is expressed *via* mRNA.

Exonuclease An enzyme that digests a nucleic acid molecule from one or both ends.

Extrachromosomal element A DNA molecule that is not part of the host cell chromosome.

Flush ends See *blunt ends*.

Foldback DNA Class of DNA which has palindromic or inverted repeat regions that re-anneal rapidly when duplex DNA is denatured.

Fusion protein A hybrid recombinant protein that contains vector-encoded amino acid residues at the N terminus.

ß-Galactosidase An enzyme encoded by the *lacZ* gene. Splits lactose into glucose and galactose.

Gamete Refers to the haploid male (sperm) and female (egg) cells that fuse to produce the diploid zygote (q.v.) during sexual reproduction.

Gel electrophoresis Technique for separating nucleic acid molecules on the basis of their movement through a gel matrix under the influence of an electric field. See *agarose* and *polyacrylamide*.

Gel retardation Method of determining protein-binding sites on DNA fragments on the basis of their reduced mobility, relative to unbound DNA, in gel electrophoresis experiments.

Gene The unit of inheritance, located on a chromosome. In molecular terms, usually taken to mean a region of DNA that encodes one function. Broadly, therefore, one gene encodes one protein.

Gene bank See *genomic library*.

Gene cloning The isolation of individual genes by generating recombinant DNA molecules, which are then propagated in a host cell which produces a clone that contains a single fragment of the target DNA.

Gene therapy The use of cloned genes in the treatment of genetically derived malfunctions *in vivo*.

Genetic code The triplet codons that determine the types of amino acid that are inserted into a polypeptide during translation. There are 64 codons for 20 amino acids, and the code is therefore referred to as *degenerate*.

Genetic fingerprinting A method which uses radioactive probes to identify bands derived from hypervariable regions of DNA (q.v.). The band pattern is unique for an individual, and can be used to establish identity or family relationships.

Genetic marker A phenotypic characteristic that can be ascribed to a particular gene.

Genome Used to describe the complete genetic complement of a virus, cell or organism.

Genomic library A collection of clones which together represent the entire genome of an organism.

Genotype The genetic constitution of an organism.

Germ line Gamete producing (reproductive) cells that give rise to eggs and sperm.

Guanine (G) Nitrogenous base found in DNA and RNA.

Heterologous Refers to gene sequences that are not identical, but show variable degrees of similarity.

Heteropolymer A polymer composed of different types of monomer. Most protein and nucleic acid molecules are heteropolymers.

Hogness box See *TATA box*.

Homologous (1) Refers to paired chromosomes in diploid organisms. (2) Used to strictly describe DNA sequences that are identical; however, the percentage homology between related sequences is sometimes quoted.

Homopolymer A polymer composed of only one type of monomer, such as polyphenylalanine (protein) or polyadenine (nucleic acid).

Host A cell used to propagate recombinant DNA molecules.

Hybrid-arrest translation Technique used to identify the protein product of a cloned gene, in which translation of its mRNA is prevented by the formation of a DNA·mRNA hybrid.

Hybrid-release translation Technique in which a particular mRNA is selected by hybridisation with its homologous, cloned DNA sequence, and is then translated to give a protein product that can be identified.

Hybridisation The joining together of artificially separated nucleic acid molecules *via* hydrogen bonding between complementary bases.

Hypervariable region (HVR) A region in a genome that is composed of a variable number of repeated sequences and is diagnostic for the individual. See *genetic fingerprinting*.

Insertion vector A bacteriophage vector that has a single cloning site into which DNA is inserted.

Intervening sequence Region in a eukaryotic gene that is not expressed *via* the processed mRNA.

Intron See *intervening sequence*.

Inverted repeat A short sequence of DNA that is repeated, usually at the ends of a longer sequence, in a reverse orientation.

In vitro Literally 'in glass', meaning in the test-tube, rather than in the cell or organism.

In vivo Literally 'in life', meaning the natural situation, within a cell or organism.

IPTG iso-propyl-thiogalactoside, a gratuitous inducer which de-represses transcription of the *lac* operon.

Kilobase (kb) 10^3 bases or base-pairs, used as a unit for measuring or specifying the length of DNA or RNA molecules.
Klenow fragment A fragment of DNA polymerase I that lacks the $5' \rightarrow 3'$ exonuclease activity.

Linker A synthetic self-complementary oligonucleotide that contains a restriction enzyme recognition site. Used to add cohesive ends (q.v.) to DNA molecules that have blunt ends (q.v.).
Locus The site at which a gene is located on a chromosome.
Lysogenic Refers to bacteriophage infection that does not cause lysis of the host cell.
Lytic Refers to bacteriophage infection that causes lysis of the host cell.

Mega (M) SI prefix, 10^6.
Messenger RNA (mRNA) The ribonucleic acid molecule transcribed from DNA that carries the codons specifying the sequence of amino acids in a protein.
Micro (μ) SI prefix, 10^{-6}.
Microinjection Introduction of DNA into the nucleus or cytoplasm of a cell by insertion of a microcapillary and direct injection.
Milli (m) SI prefix, 10^{-3}.
Monocistronic Refers to an RNA molecule encoding one function.
Monomer The unit that makes up a polymer. Nucleotides and amino acids are the monomers for nucleic acids and proteins respectively.
Mosaic An embryo or organism in which not all the cells carry identical genomes.
Multiple cloning site (MCS) A short region of DNA in a vector that has recognition sites for several restriction enzymes.
Mutagenesis The process of inducing mutations in DNA.
Mutant An organism (or gene) carrying a genetic mutation.
Mutation An alteration to the sequence of bases in DNA. May be caused by insertion, deletion or modification of bases.

Nano (n) SI prefix, 10^{-9}.
Native protein A recombinant protein that is synthesised from its own N terminus, rather than from an N terminus supplied by the cloning vector.
Nested fragments A series of nucleic acid fragments that differ from each other (in terms of length) by one or only a few nucleotides.
Nick translation Method for labelling DNA with radioactive dNTPs.
Northern blotting Transfer of RNA molecules onto membranes for the detection of specific sequences by hybridisation.
N terminus Amino terminus, defined by the $-NH_2$ group of an amino acid or protein.
Nuclease An enzyme that hydrolyses phosphodiester bonds.
Nucleoid Region of a bacterial cell in which the genetic material is located.

Nucleoside A nitrogenous base bound to a sugar.

Nucleotide A nucleoside bound to a phosphate group.

Nucleus Membrane-bound region in a eukaryotic cell that contains the genetic material.

Oligo Prefix meaning few, as in oligonucleotide or oligopeptide.

Oligo(dT)-cellulose Short sequence of deoxythymidine residues linked to a cellulose matrix, used in the purification of eukaryotic mRNA.

Oligolabelling See *primer extension*.

Oligomer General term for a short sequence of monomers.

Oligonucleotide A short sequence of nucleotides.

Oligonucleotide-directed mutagenesis Process by which a defined alteration is made to DNA using a synthetic oligonucleotide.

Operator Region of an operon, close to the promoter, to which a repressor protein binds.

Operon A cluster of bacterial genes under the control of a single regulatory region.

Palindrome A DNA sequence that reads the same on both strands when read in the same (e.g. $5' \rightarrow 3'$) direction. Examples include many restriction enzyme recognition sites.

Pedigree analysis Determination of the transmission characteristics of a particular gene by examination of family histories.

α-Peptide Part of the ß-galactosidase protein, encoded by the *lacZ'* gene fragment.

Phage See *bacteriophage*.

Phasmid A vector containing plasmid and phage sequences.

Phenotype The observable characteristics of an organism, determined both by its genotype (q.v.) and its environment.

Phosphodiester bond A bond formed between the 5'-phosphate and the 3'-hydroxyl groups of two nucleotides.

Pico (p) SI prefix, 10^{-12}.

Plaque A cleared area on a bacterial lawn caused by infection by a lytic bacteriophage.

Plasmid A circular extrachromosomal element found naturally in bacteria and some other organisms. Engineered plasmids are used extensively as vectors for cloning.

Polyacrylamide A cross-linked matrix for gel electrophoresis (q.v.) of small fragments of nucleic acids, primarily used for electrophoresis of DNA. Also used for electrophoresis of proteins.

Polyadenylic acid A string of adenine residues. Poly(A) tails are found at the 3' ends of most eukaryotic mRNA molecules.

Polycistronic Refers to an RNA molecule encoding more that one function. Many bacterial operons are expressed *via* polycistronic mRNAs.

Polygenic trait A trait determined by the interaction of more than one gene, e.g. eye colour in humans.

Polylinker See *multiple cloning site.*

Polymer A long sequence of monomers.

Polymerase An enzyme that synthesises a copy of a nucleic acid.

Polymerase chain reaction (PCR) A method for the selective amplification of DNA sequences.

Polynucleotide A polymer made up of nucleotide monomers.

Polynucleotide kinase (PNK) An enzyme that catalyses the transfer of a phosphate group onto a 5′-hydroxyl group.

Polystuffer An expendable stuffer fragment in a vector that is composed of many repeated sequences.

Pribnow box Sequence found in prokaryotic promoters that is required for transcription initiation. The consensus sequence (q.v.) is TATAAT.

Primary transcript The initial, and often very large, product of transcription of a eukaryotic gene. Subjected to processing to produce the mature mRNA molecule.

Primer extension Synthesis of a copy of a nucleic acid from a primer. Used in labelling DNA and in determining the start site of transcription.

Probe A labelled molecule used in hybridisation procedures.

Prokaryotic The property of lacking a membrane-bound nucleus, e.g. bacteria such as *E. coli.*

Promoter DNA sequence(s) lying upstream from a gene, to which RNA polymerase binds.

Pronucleus One of the nuclei in a fertilised egg prior to fusion of the gametes.

Prophage A bacteriophage maintained in the lysogenic state in a cell.

Protein A condensation heteropolymer composed of amino acid residues.

Protoplast A cell from which the cell wall has been removed.

Prototroph A cell that can grow in an unsupplemented growth medium.

Purine A double-ring nitrogenous base such as adenine and guanine.

Pyrimidine A single-ring nitrogenous base such as cytosine, thymine and uracil.

Recombinant DNA A DNA molecule made up of sequences that are not normally joined together.

Regulatory gene A gene that exerts its effect by controlling the expression of another gene.

Repetitive sequence A sequence that is repeated a number of times in the genome.

Replacement vector A bacteriophage vector in which the cloning sites are arranged in pairs, so that the section of the genome between these sites can be replaced with insert DNA.

Replication Copying the genetic material during the cell cycle. Also refers to the synthesis of new phage DNA during phage multiplication.

Replicon A piece of DNA carrying an origin of replication.

Restriction enzyme An endonuclease that cuts DNA at sites defined by its recognition sequence.

Restriction fragment A piece of DNA produced by digestion with a restriction enzyme.

Restriction fragment length polymorphism (RFLP) A variation in the locations of restriction sites bounding a particular region of DNA, such that the fragment defined by the restriction sites may be of different lengths in different individuals.

Restriction mapping Technique used to determine the location of restriction sites in a DNA molecule.

Retrovirus A virus that has an RNA genome that is copied into DNA during the infection.

Reverse transcriptase An RNA-dependent DNA polymerase found in retroviruses, used *in vitro* for the synthesis of cDNA.

Ribonuclease (RNase) An enzyme that hydrolyses RNA.

Ribonucleic acid (RNA) A condensation heteropolymer composed of ribonucleotides.

Ribosomal RNA (rRNA) RNA that is part of the structure of ribosomes.

Ribosome The 'jig' that is the site of protein synthesis. Composed of rRNA and proteins.

Ribosome-binding site A region on an mRNA molecule that is involved in the binding of ribosomes during translation.

RNA processing The formation of functional RNA from a primary transcript (q.v.). In mRNA production this involves removal of introns, addition of a 5′ cap and polyadenylation.

S_1 mapping Technique for determining the start point of transcription.

S_1 nuclease An enzyme that hydrolyses (degrades) single-stranded DNA.

Scintillation counter A machine for determining the amount of radioactivity in a sample.

Screening Identification of a clone in a genomic or cDNA library (q.v.) by using a method that discriminates between different clones.

Selection Exploitation of the genetics of a recombinant organism to enable desirable, recombinant genomes to be selected over non-recombinants during growth.

Shine–Dalgarno sequence See *ribosome-binding site*.

Site-directed mutagenesis See *oligonucleotide-directed mutagenesis*.

Southern blotting Method for transferring DNA fragments onto a membrane for detection of specific sequences by hybridisation.

Specific activity The amount of radioactivity per unit material, e.g. a labelled probe might have a specific activity of 10^6 counts/minute per microgram. Also used to quantify the activity of an enzyme.

Sticky ends See *cohesive ends*.

Structural gene A gene that encodes a protein product.

Stuffer fragment The section in a replacement vector (q.v.) that is removed and replaced with insert DNA. See *polystuffer*.

Substitution vector See *replacement vector*.

Tandem repeat A repeat composed of an array of sequences repeated contiguously in the same orientation.

TATA box Sequence found in eukaryotic promoters. Also known as the *Hogness* box, it is similar to the Pribnow box (q.v.) found in prokaryotes, and has the consensus sequence TATAAAT.

Temperate Refers to bacteriophages that can undergo lysogenic infection of the host cell.

Terminal transferase An enzyme that adds nucleotide residues to the 3′ terminus of an oligo- or polynucleotide.

Tetracycline (Tc) A commonly used antibiotic.

Thymine (T) Nitrogenous base found in DNA only.

***Trans*-acting element** A genetic element that can exert its effect without having to be on the same molecule as a target sequence. Usually such an element encodes a protein product (perhaps an enzyme or a regulatory protein) that can diffuse to the site of action.

Transcription (T_C) The synthesis of RNA from a DNA template.

Transcriptional unit The DNA sequence that encodes the RNA molecule, i.e. from the transcription start site to the stop site.

Transfection Introduction of purified phage or virus DNA into cells.

Transfer RNA (tRNA) A small RNA ($\sim$75–85 bases) that carries the anticodon and the amino acid residue required for protein synthesis.

Transformant A cell that has been transformed by exogenous DNA.

Transformation The process of introducing DNA (usually plasmid DNA) into cells.

Transgenic An organism that carries DNA sequences that it would not normally have in its genome.

Translation (T_L) The synthesis of protein from an mRNA template.

Transposable element A genetic element that carries the information that allows it to integrate at various sites in the genome. Transposable elements are sometimes called 'jumping genes'.

Uracil (U) Nitrogenous base found in RNA only.

Vector A DNA molecule that is capable of replication in a host organism, and can act as a carrier molecule for the construction of recombinant DNA.

Virulent Refers to bacteriophages that cause lysis of the host cell.

Virus An infectious agent that cannot replicate without a host cell.

Western blotting Transfer of electrophoretically separated proteins onto a membrane for probing with antibody.

X-gal 5-bromo-4-chloro-3-indolyl-ß-D-galactopyranoside: a chromogenic substrate for ß-galactosidase; on cleavage it yields a blue-coloured product.

YAC Yeast artificial chromosome, a vector for cloning very large pieces of DNA in yeast.

Zygote Single-celled product of the fusion of a male and a female gamete (q.v.). Develops into an embryo by successive mitotic divisions.

Index

Page numbers followed by F refer to Figures;
page numbers followed by T refer to Tables.